Environmental Division 2016

Core Programming Area at the 2016 AIChE Spring Meeting and 12th Global Congress on Process Safety

Houston, Texas, USA
10-14 April 2016

ISBN: 978-1-5108-2493-5

Printed from e-media with permission by:

Curran Associates, Inc.
57 Morehouse Lane
Red Hook, NY 12571

Some format issues inherent in the e-media version may also appear in this print version.

Printed by Curran Associates, Inc. (2016)

AIChE
120 Wall Street, FL 23
New York, NY 10005-4020

Phone: (800) 242-4363
Fax: (203) 775-5177

www.aiche.org

Additional copies of this publication are available from:

Curran Associates, Inc.
57 Morehouse Lane
Red Hook, NY 12571 USA
Phone: 845-758-0400
Fax: 845-758-2633
Email: curran@proceedings.com
Web: www.proceedings.com

TABLE OF CONTENTS

443215 WSA (Wet gas Sulfur Acid) Technology - A Competitive Solution for Sulfur Management

Abstract:

For more than 30 years Haldor Topsoe has designed plants for purification of sulfur containing waste gases derived from a wide variety of process industries. The WSA (Wet gas Sulfuric Acid) technology is efficient for recovering H_2S or other sulfur compounds in the form of commercial grade sulfuric acid. Moreover, the exothermic reaction energy generated when converting the sulfurous off-gas is efficiently recovered in the process as valuable superheated high pressure steam available for export.

The WSA technology is also characterized by high sulfur removal efficiency with more than 99% sulfur recovery, up to 99.99% when required, and is always in compliance with emission legislation. The process does not consume any chemicals nor will there be any solid or liquid effluents requiring additional treatment. This in combination with very low cooling water consumption makes WSA technology a very competitive solution for treatment of acid gases.

The WSA technology has proven feasible for treating sulfurous gases in many different applications, and today more than 140 units have been contracted worldwide handling a wide range of gas flows and with daily sulfuric acid production ranging from 4 to 1,140 MTPD.

The paper will explain the basic principles of the WSA technology and will include economical figures and the comparison with the typical alternatives for sulfur removal.

443717 Air-Quality Conscious Scheduling for Multiple Olefin Plant Start-Ups

Abstract:

The ground-level ozone is a pervasive air pollutant, which can be formed from flare emissions during chemical plant start-up operations. Especially in the concentrated industrial areas, the regional air quality (i.e., the ozone concentration) might be aggravated by simultaneously start-up emissions from multiple chemical plants. Thus, it is environmentally important for optimal scheduling of multi-plant start-ups under their manufacturing allowances to minimize the possible adverse air-quality impact. A systematic methodology on air-quality conscious scheduling for multi-plant start-ups has been developed. It demonstrates that multi-plant start-ups without any coordination could result in the significant air-quality impact. However, an optimal scheduling plan with several-hour tuning on the starting time of their start-up operations would significantly reduce about 80 % ozone increment. The study couples process dynamic simulation for industrial flare emissions with regional air-quality modeling together. It can provide valuable and quantitative supports for all relevant stakeholders, including environmental agencies, regional plants, and local communities.

444021 Integrated Waste Water Collection and Treatment System for Zero Waste Discharge Liquid Plant

Abstract:

Efficient design for collection of waste and effluent treatment is major challenge in EPC industry today. In this paper two cases of integration of waste water collection & treatment system are discussed. In first case, Power plant is considered and in other case LNG plant is taken. The sizing of waste water system and treatment of the effluents for zero waste discharge are compared for both of the cases. This paper throws light on significant savings by rightly sizing the collection and treatment system which majorly includes sumps, sump pumps, oil water separator, equalization tank, clarified water tank, and clarifier based on the intermittent and continuous waste from the plant. The flow sheets have been designed for power plant as well as LNG plant for integrated waste water collection & treatment system for zero waste discharge liquid plant. All the discharges like oil, waste water and chemicals are treated effectively and can be regenerated to use in the same plant. It is not only cost effective but also environment friendly.

BURNS & McDONNELL

The Refinery "RTR" Rule – Major Refinery Impacts and Targeted Solutions

2016 AIChE Spring Meeting
Houston, TX
April 11, 2016

Blake Soyars, Section Manager, Environmental

Mark W. Lockhart, PE, Process Technology Manager Process & Industrial

Jim Corbit, Department Manager, Environmental

Greg Neve, Project Manager, Process & Industrial

Agenda

- What is the "RTR"
- "RTR" – Numerous Potential Impacts
- Long-Lead Compliance Projects Timing
- Roadmap to Success
- FCCU
  - Potential Effects and Solutions
- Delayed Coker:
  - Potential Effects and Solutions
- Flare Systems:
  - Potential Effects and Solutions
- Catalytic Reforming
  - Potential Effects and Solutions
- Benzene Fenceline Monitoring
- Conclusions

BURNS & McDONNELL 2

What is the "RTR"?

- "Risk & Technology Review" rulemaking only applies to petroleum refineries
- New US EPA Clean Air Act rulemaking finalized Dec. 1, 2015
- Five key rules updated
  - NSPS J & NSPS Ja focus on criteria pollutants VOC, SOx, NOx, CO & PM
  - MACT Y, MACT CC & MACT UUU focus on Hazardous Air Pollutants (HAP)
- Almost every US refinery will be affected
- Complex new requirements (following years of actual emissions reductions)
- Potential high initial compliance costs for many refiners:
  - EPA estimated **$2 million** per refinery average across 143 US refineries
  - American Petroleum Institute estimated **$8 million** per refinery average
  - Actual costs will depend on many site-specific factors
- Most costs will occur in 2016 - 2018 for overall Jan. 30, 2019 deadline

Refinery Risk & Technology Review (RTR): numerous potential impacts

- New/revised air emission standards drive compliance projects & add new requirements to certain economic projects.
- Each refinery needs detailed technical evaluations to identify, evaluate & address potential RTR impacts, including:
  - **FCCUs** – do regen pollution controls meet new requirements at all times?
  - **Reformers** – do formerly "exempted" regen vents <5 psig meet new requirements?
  - **Delayed Cokers** – are modifications needed for >2 psig decoking vents to atmosphere?
  - **Flares** – flare gas recovery and/or control/instrumentation upgrades needed?
  - **Storage Tanks** – new floating roof retrofits or roof fitting modifications required?
  - **Startup/Shutdown** – do emissions comply after RTR eliminates SSM exemptions?
  - **Overall** – does refinery have roadmap to evaluate all impacts and meet all deadlines?

RTR Timeline: long-lead compliance projects may be short on time

- For existing refinery sources, comply "as expeditiously as practical, but no later than..."
  - **Startup/Shutdown/Malfunction** – "exemptions" eliminated on Jan. 30, 2016 & replaced by "general duty" requirements, then new SSM limits apply by Aug. 1, 2017
  - **Storage tank standards** – implement by April 29, 2016 (can request more time)
  - **FCC emissions testing** – new tests/frequency with first new tests by August 1, 2017
  - **Benzene fenceline monitoring** – start by Jan. 30, 2018
  - **All other requirements** – comply by Jan. 30, 2019
- Will EPA revise the rule? minor "corrections" were proposed & industry will request more changes (outcomes uncertain)

BURNS & McDONNELL 5

Fluid Catalytic Cracking Units: potential project impacts for normal operation

- **RTR focus** for FCCUs is on the catalyst regeneration vent, which involves complex equipment that is extremely expensive to upgrade or modify
- **RTR impacts** are highly dependent on unit-specific configuration & operation: many units are ready to comply but detailed review needed for each unit
- **Normal Operating Requirements**:
  - Previous limits & requirements were selectively made more stringent
  - New testing by Aug, 1, 2017 & most other requirements apply Jan. 30, 2019
- **Startup, Shutdown & Hot Standby Requirements:**
  - Previous SSM "exemptions" replaced with new requirements matrix
  - Examples: new >20 ft/sec cyclone velocity for metal HAP & new 1% excess O2 limit for organic HAP
  - Some changes effective immediately: complete transition by Aug. 1, 2017

BURNS & McDONNELL 6

Fluid Catalytic Cracking Units: Solutions

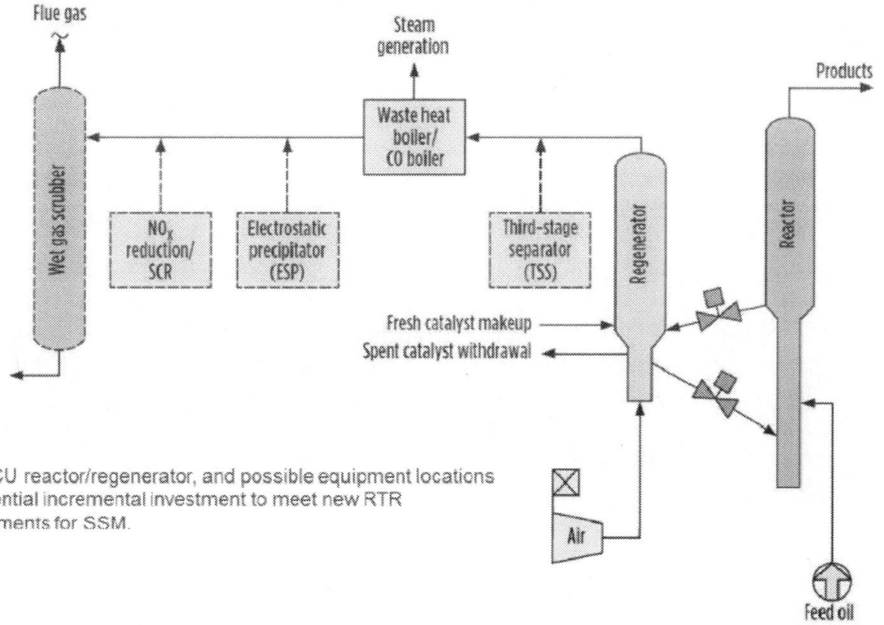

An FCCU reactor/regenerator, and possible equipment locations for potential incremental investment to meet new RTR requirements for SSM.

BURNS & McDONNELL 7

Fluid Catalytic Cracking Units – SSM

Third stage separators (TSS)	CO boiler (Partial burn)	Electrostatic precipitator (ESP)	Selective catalytic reduction (SCR)	Wet gas scrubber (WGS)	Comments:
				X	WGS assists with meeting PM, NO_x and SO_x requirements. CO met by complete combustion operation. Robust system.
	X				CO boiler assists with meeting CO. Evaluate incremental investment for TSS or ESP for meeting PM requirements on startup. SCR can assist with meeting NO_x.
		X			ESP assists in meeting PM but typically bypassed on startup. CO, NO_x and SO_x controlled by operating conditions in FCCU.
	X			X	CO boiler assists with meeting CO. WGS assists with meeting PM, NO_x and SO_x. Robust system.

Equipment exists (dark shading)

Equipment to consider as a minimum incremental investment (light shading)

BURNS & McDONNELL 8

Delayed Coking Units: potential project impacts

- **Decoking vent pressure standards tightened for all delayed cokers:**
  - Old rule allowed drum venting as soon as pressure dropped to <5 psig outlet
  - New rule requires drum vapors to be processed or treated for organic HAP emission reduction down to <2 psig outlet
  - <2 psig limit applies to venting, draining, deheading & cutting
  - Additional technical requirement differences for "new" units versus "existing" units and for water overflow units versus other unit types
- **Emission reduction requirements unchanged:** all vapors that cannot meet <2 psig exemption must be recovered or treated to existing standards:
  - route to flare; or
  - reduce organic HAP by at least 98%; or
  - meet <20 ppmv Total Organic Carbon

Delayed Coking Units - Solutions

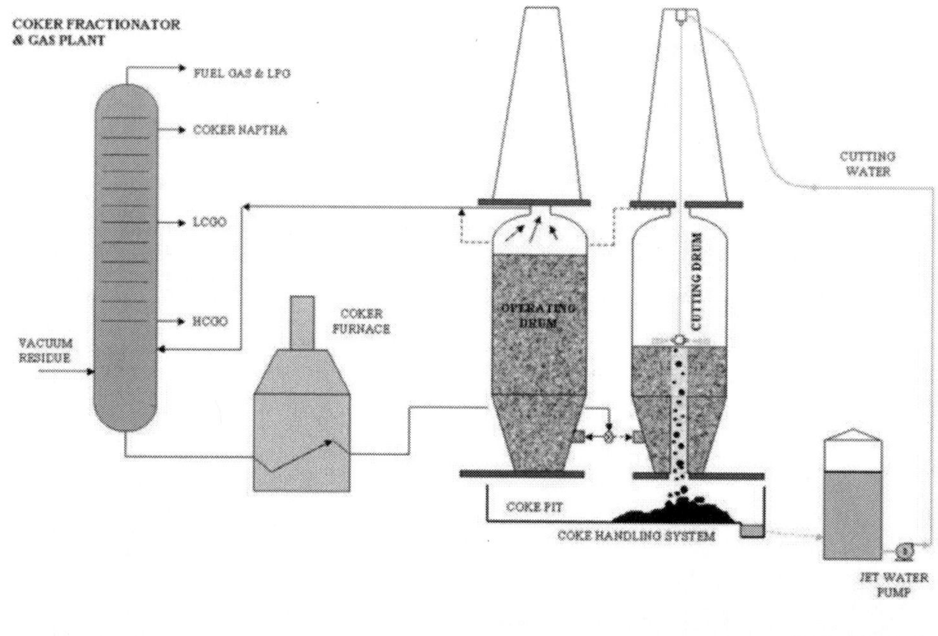

Source: Adams Project Managers, Inc.

Potential Mitigations

- **Achieving < 2 Psig Limit**
  - Use steam ejectors on blowdown drum
  - Use compressors on blowdown drum
  - Use compressors on coke drum overhead
  - Reduce pressure drop through blowdown system
  - Water filling in coke drums
  - Flare gas recovery compressor

Delayed Coking Units - Solutions

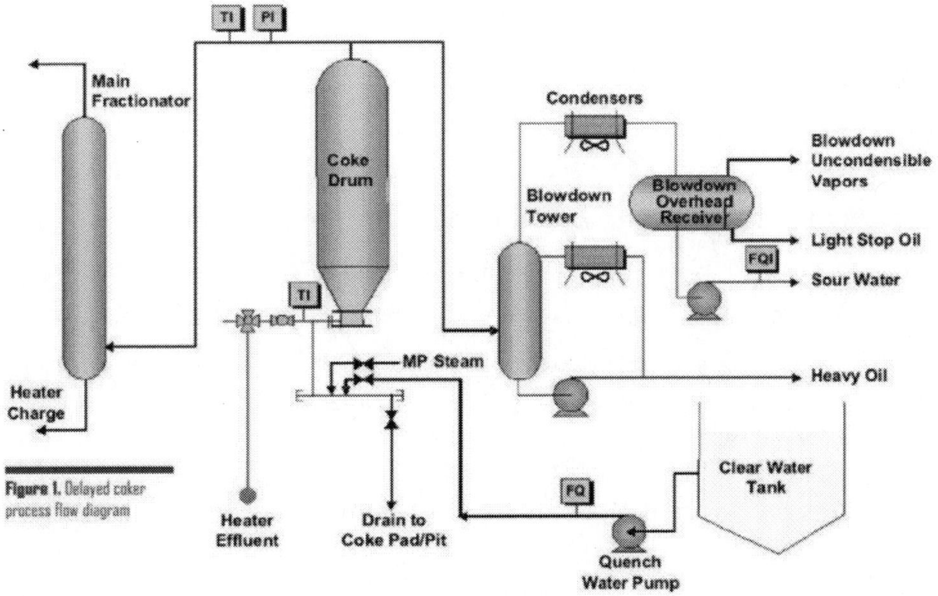

Figure 1. Delayed coker process flow diagram

Source: Lummus Technology, USA.

Design Considerations

- Can Coke Drums now achieve < 2 psig?
- Does achieving < 2 psig require more time which may affect capacity?
- What is the pressure drop through the blowdown system?
- Review increasing blowdown condensing capacity
- If add more motive force:
  - Ejectors – motive fluid; steam availability?
  - Compressors – Electrical capacity
- Main Fractionator – overhead cooling, compression and sour water handling.
- Consider a Flare Gas Recovery (FGR) system

BURNS & McDONNELL 13

Flare Systems: potential project impacts

- **RTR overhauls previous flare requirements**
- **New flare gas measurements required:**
  - Continuous volume flow measurement required for all "Combustion Zone" (CZ) contributing flows: vent gas, sweep, total steam, pre-mix, etc. (not pilot)
  - Quality of all flared gasses must be analyzed by instruments or sampling
- **Revised operating constraints for "normal" flare operation:**
  - New minimum Net Heating Value (NHV) ≥270 Btu/scf limit applies to all CZ flows
  - Tip velocity limits are similar to pre-RTR rules but apply to all CZ flows
  - NHV & velocity compliance calculations now required every 15-minute block
  - Startup/shutdown, excess fuel gas, PRV/PSV malfunctions, etc. regulated as "normal"
- **Emergency & high rates** >smokeless capacity subject to detailed new requirements
- **Bottom line**: RTR changes will drive detailed analysis & many flare projects

BURNS & McDONNELL 14

Flare System Solutions

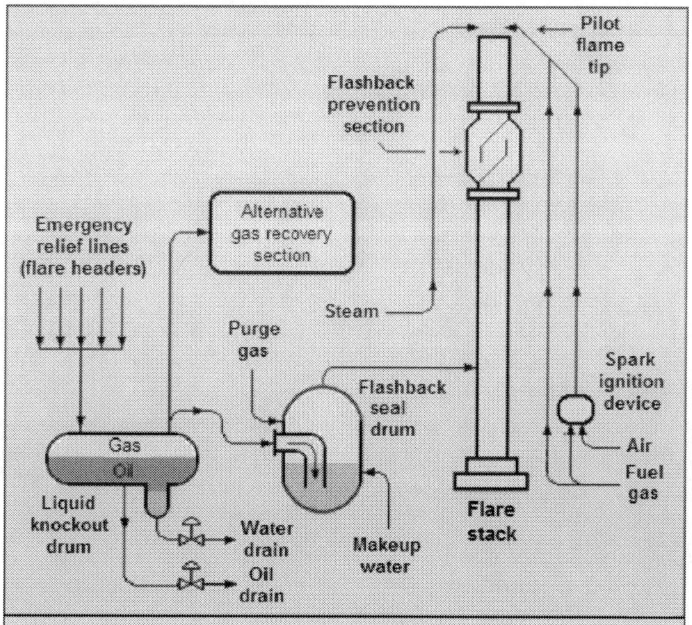

Diagram of an overall flare stack system in a petroleum refinery

Source: chemengineering.wikispaces.com.

BURNS & McDONNELL 15

Catalytic Reforming Units: potential project impacts

- **Scope of RTR changes for reformers:**
  - RTR only changes organic HAP requirements before catalyst burn-off
  - Inorganic HAP requirements for burn-off are unaffected by RTR
- **Old <5 psig vent "exemption" narrowed with more units regulated:**
  - Exemption now "clarified" to only apply to passive vessel depressurization
  - Any active purge vent (even if <5 psig) with sufficient pressure to route into emission treatment must now be treated to remove organic HAP
- **Organic HAP standards unchanged:** previously "exempted" <5 psig vents can no longer be vented unless routed to control: flare or other device to meet 98% HAP reduction or <20 ppmv TOC

BURNS & McDONNELL 16

Catalytic Reforming

Continuous Catalytic Regeneration (CCR) Units

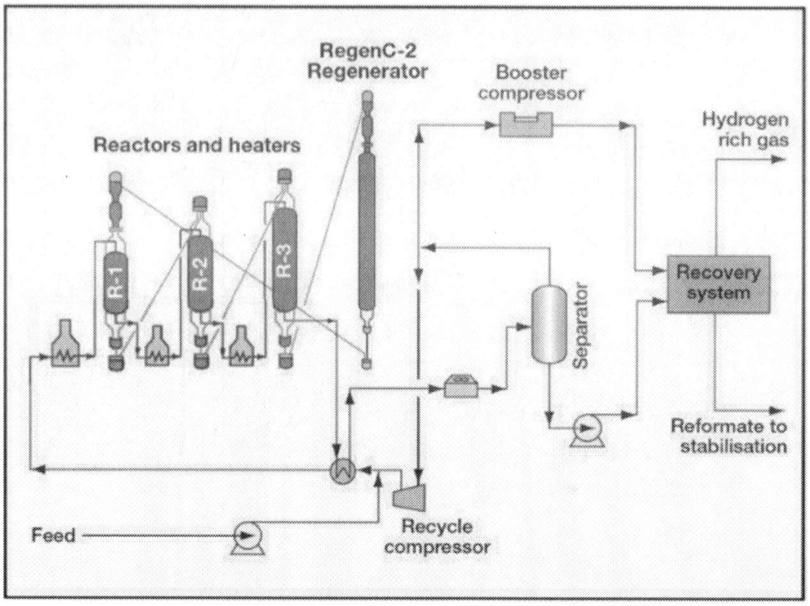

Source: Axens IFP Group Technologies.

Catalytic Reforming

Semi-Regenerative Units

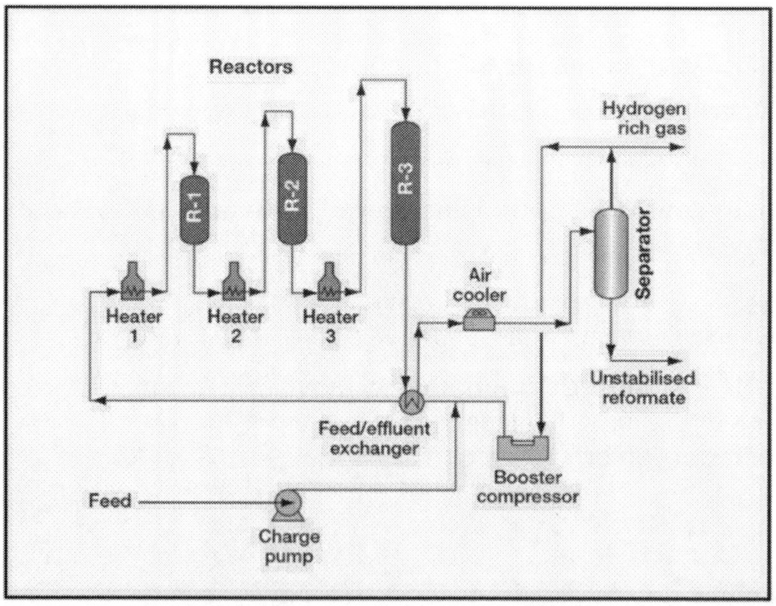

Source: Axens IFP Group Technologies.

Benzene Fenceline Monitoring Requirements

New program required to measure ambient benzene around refinery:

- Install 12 to 24 permanent fenceline/perimeter monitors (depending on refinery footprint) plus any background monitors to identify off-site benzene contributions
- Passive absorbent technology detects very low concentrations 24/7
- Meteorological station also required but can use existing/nearby met station
- Benzene data must be disclosed to public with associated new community perception & activist empowerment challenges
- Conduct root cause analysis if annual average measured benzene is >9 µg/m3 after discounting off-site contributions
- Multiple high measurements with related causes may lead to project requirements or increased compliance costs e.g., enhanced piping leak detection and repair program

BURNS & McDONNELL

19

Benzene Fenceline Monitoring Example

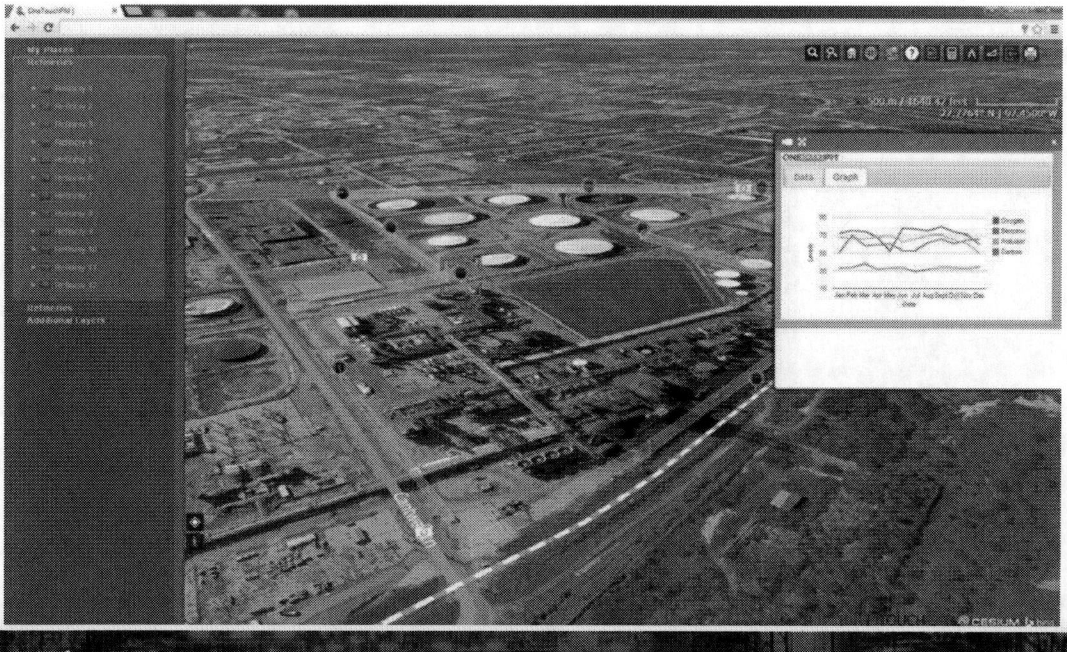

BURNS & McDONNELL

20

Other Impacts

- **Sulfur Plant**: new S/U & S/D requirements for purging gas
- **Storage Tanks**: some existing tanks may need new floating roofs, some existing floating roof fittings may need upgrades with April 29, 2016 deadline
- **Relief Valves:** new equipment requirements so that every release has: immediate detection, time/duration measurement, and operator notification
- **Maintenance Vapor Clearing:** new requirements apply even to small equipment clearing: affects procedures & compliance documentation
- **Onstream Analyzer Vents**: are now regulated process vents requiring vent-specific evaluations (many should meet flow/concentration limits)
- **Marine Vessel Loading**: certain loading was exempt from submerged-fill requirements but those exemptions were eliminated

Floating Roof Tanks, and Other Impacts:

External Floating Roof Tank (Pontoon Type)

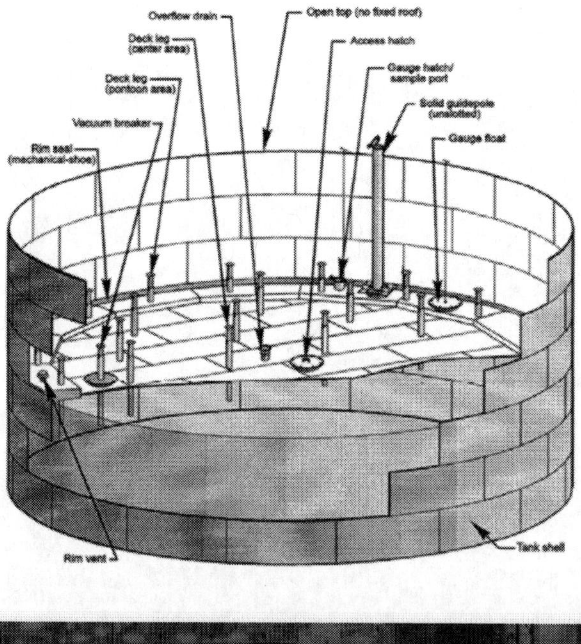

Source: US EPA, AP 42, November 2006.

Conclusions/Recommendations

- **Complexity:** is not reflected in slides, see paper for more detail
- **Priority**: plan for significant RTR compliance implementation costs with typical costs estimated in $2-8 million range per refinery
- **Scope**: the RTR can impact key refinery units and drive complex projects that may need coordination with turnarounds or other projects
- **Urgency**: the Jan 2019 overall compliance deadline combines with earlier deadlines for certain individual requirements to create schedule difficulty
- **Roadmap** is recommended to reflect priority, scope & urgency to chart course for successful compliance implementation with managed impacts
- **Team**: a purpose-built RTR implementation team is recommended to engage many necessary disciplines & stakeholders
- **Success** depends on site-specific RTR roadmap, implemented by an integrated RTR team, aligned with commercial priorities, and driven by compliance commitment

BURNS & McDONNELL 23

Contact Information

Blake Soyars, PE - Author/Presenter

Austin | 512-872-7150 | brsoyars@burnsmcd.com

Mark Lockhart, PE - Author/Presenter

Houston | 832-214-1975 | mlockhart@burnsmcd.com

Jim Corbit - Author

Houston | 832-214-2818 | jrcorbit@burnsmcd.com

Greg Neve - Author

Houston | 832-389-5746 | gneve@burnsmcd.com

BURNS & McDONNELL

CREATE AMAZING.

16AIChE
Spring Meeting, Houston, TX

The Principle of Intrinsic Reality Solving the World's Woes through Chemical Engineering and Economics

Jed Moyano Bellen, SCPM, PMP
Hyundai Engineering & Construction Co. Ltd.
9 Temasek Blvd., #27-02 Suntec Tower 2, Singapore 038989
jed.bellen@hdec.co.kr

Prepared for Presentation at
American Institute of Chemical Engineers
2016 Spring Meeting
11th Global Congress on Process Safety
Houston, Texas
April 10-14, 2016

UNPUBLISHED

The Principle of Intrinsic Reality Solving the World's Woes through Chemical Engineering and Economics

Jed Moyano Bellen, SCPM, PMP
Hyundai Engineering & Construction Co. Ltd.
9 Temasek Blvd., #27-02 Suntec Tower 2, Singapore 038989
jed.bellen@hdec.co.kr

Keywords: Principle of Intrinsic Reality, Chemical Engineering, Economics, Sustainability

Abstract

Human beings in profound rationality always seek for explanations of the phenomena around them. To understand these phenomena, human beings have developed systematized body of knowledge as a practical way of dealing with life's challenges. Among the systematized body of knowledge developed are chemical engineering and economics.

An avant-garde way of looking at each of these disciplines leads to the Principle of Intrinsic Reality. This principle relates the fundamental laws of the two disciplines to glean insights on the constraints imposed by conservation laws on economic systems at both the macro and micro levels.

It can be argued that if mass and energy in the universe are constant, then logically diminishing returns and marginal utility are bound to be observed. This logic is quite evident in a microorganism culture which follows the Monod growth model.

It can also be thought that conservation laws imply scarcity of resources. As such, it enjoins everyone to find optimal and efficient use of resources in the context of trade-offs, production decisions and sustainability. It even calls for the development of new technologies to produce more outputs from less inputs (i.e. shifting of production possibility frontier to the right). In this way, the Principle of Intrinsic Reality leads to the imperative for sustainable optimization.

Furthermore, it is deemed that the conservation laws can explain the phenomenon called "polarity of wealth." As mass and energy when harnessed create wealth, then the constancy of mass and energy implies boundary for wealth creation. In effect, wealth can neither be created nor destroyed but can only shift from one entity to another thereby creating polarity in wealth distribution. This assertion can be the key towards more equitable restructuring of economic systems to close income gap and thereby result to stable and resilient populace.

In conclusion, this paper envisions for a paradigm shift from traditional money-centric view of chemical engineering and economics to a synergistic convergence of disciplines which can powerfully tackle the world's woes and thereby effect sustainability.

1. The World in which We Live Today

The past few years saw unprecedented global challenges which have put into question existing economic models and scientific paradigms. Among these challenges are the dystopian global economy, climate change, environmental degradation and terrorism. The magnitude of these challenges has grown to a profound level that newer ways of thinking and perspectives become necessary. With that end in mind, this paper attempts to look at the fundamental laws of chemical engineering and economics in new ways to recognize our intrinsic reality from which these pressing challenges can be tackled at large.

2. The Principle of Intrinsic Reality

It is an observable fact that advancement in science increasingly blurs boundaries of various disciplines in tackling proactively the pressing global issues or problems. Within this purview, two seemingly diverging disciplines namely chemical engineering and economics can be taken together to glean insights on our intrinsic reality and what it means for our economic and ecological systems on a larger scale.

In the author's contention, intrinsic reality is grounded on the fundamental laws of chemical engineering and economics. These laws are the Conservation Laws (Chemical Engineering) and Law of Diminishing Returns (Economics). It can be argued that the Law of Diminishing Returns is observable because of the constraints imposed by the Conservation Laws. As such, this leads to the Principle of Intrinsic Reality which can be stated thus;

> *In any system, both at the macro- and micro-levels, the limiting factor for growth is the constancy of potential inputs.*

Potential inputs mean not the actual inputs to processes but the constant material or energy available which can be harnessed to create wealth. The reasoning or proof of this principle is presented in subsequent discussions.

2.1 Proof Based on the Monod Growth Model

The best illustration to understand the Principle of Intrinsic Reality is the Monod Growth Model. Monod in his 1949 seminal paper presented an empirical relation between exponential growth rate and concentration of essential nutrient. The data was fitted into hyperbolic equation:

$$R = R_K\left(\frac{C}{C_1 + C}\right)$$

In this equation, R is the growth rate during the exponential phase; R_K is the rate limit for increasing concentrations of C; C is the concentration of the nutrient; and C_1 is the concentration of nutrient at which the rate is half the maximum.

Figure 1: Graphical Representation of Monod Growth Model

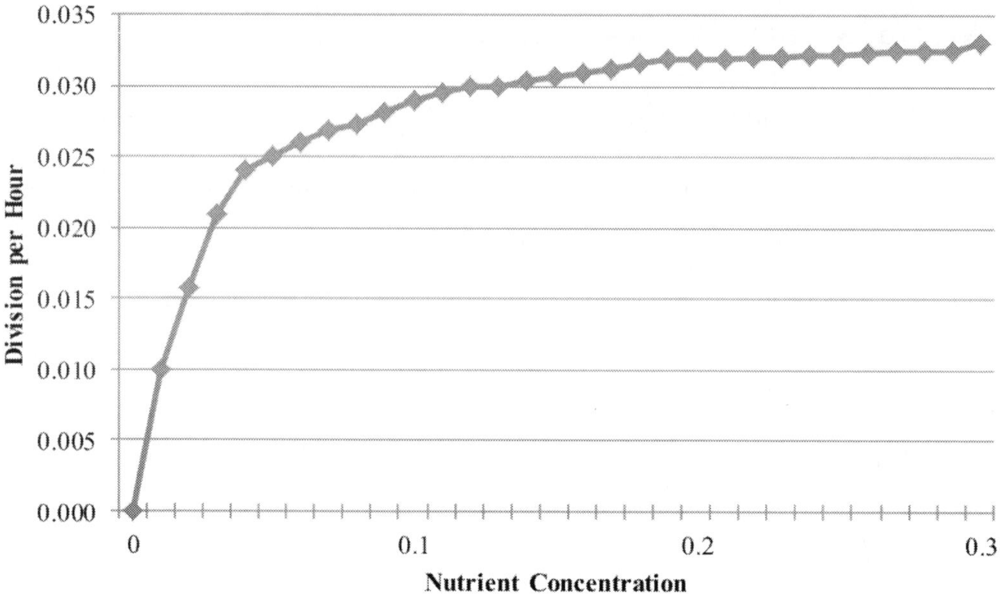

Monod (1949) further pointed out that the limiting factors commonly found are exhaustion of nutrients, accumulation of toxic metabolic products and changes in ion equilibrium.

The foregoing discussion has been observed at the micro-level with the bacterial culture. At the macro-level (i.e. the world at large), analogous phenomena are also observed to occur because of the same constraint imposed by the constant quantity of resources (i.e. material and energy). One example of these phenomena is that human beings cannot grow indefinitely. If they do, then how can the world sustain that growth with limited available resources (e.g. foods) to support it?

Another analogous phenomenon is in population growth. Thomas Malthus (1789) predicted that when population growth exceeds agricultural production, the population level is forced to return to subsistence-level conditions. This phenomenon as Malthus (1789) believed is due to the restriction of available resources supporting population growth, which others have claimed to be exploding as the earth becomes over populated.

In a surprising manner, a similar phenomenon can be observed in production processes in which the increase in a single factor of production assuming all other factors constant, results not to more marginal output but to a reduced one. This was attributed by classical economists, Thomas Malthus and David Ricardo, to the decreasing quality of inputs while neoclassical economists attributed this to the disruption in the entire production process as additional units of input are added to a fixed amount of capital. But in view of the Principle of Intrinsic Reality, the phenomenon is viewed intrinsic to the production system in which case infinite amounts of mass or energy cannot be processed with perfect efficiency. Hence, the law of conservation of mass and energy comes into the fore.

It can be deduced from the discussion that there is an underpinning intrinsic reality to the more obvious phenomena in the world from the micro-scale (bacterial growth) to the macro-scale (human growth and production processes).

2.2 Proof Based on Diminishing Marginal Utility

Another way to prove the Principle of Intrinsic Reality is through the diminishing marginal utility observable among humans and perhaps other organisms as well. Marginal utility, as defined in economics, is the gain from an increase or the loss from a decrease in the consumption of goods or services. The diminishing marginal utility is an observable phenomenon in which the utility or satisfaction an individual gains diminishes as he or she consumes more of the same unit of goods or services. Graphically, this can be represented as in Figure 2.

Figure 2: Graphical Representation of Diminishing Marginal Utility

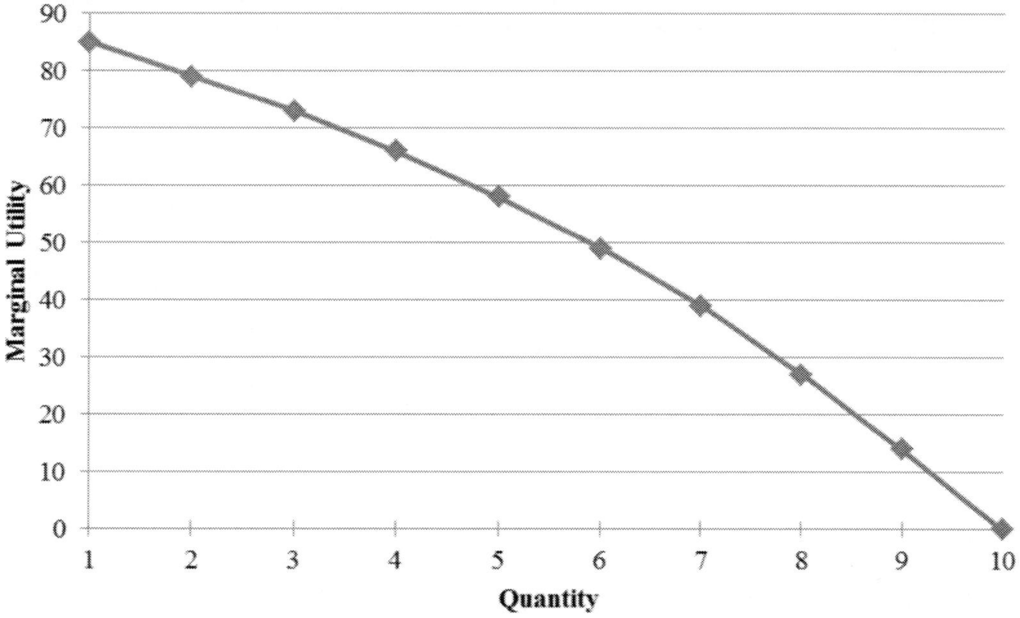

It can be deduced from this phenomenon and in support of the discussion in section 2.1 that human beings cannot grow indefinitely due to the fact that in our intrinsic reality, there is always a limit so as the constancy of available resources (i.e. mass and energy) can support our existence.

This means that in view of our intrinsic reality, we cannot have infinite greed otherwise an upsurge of unintended consequences are bound to happen (i.e. environmental degradation, terrorism and eventually demise of humanity).

Therefore, recognizing our intrinsic reality is the first step towards solving the world's woes and thereby resulting to sustainability and security.

2.3 Expanding the Production Possibility Curve through R&D

The caveat here is that the constant mass or energy (i.e. the limits) leads to trade-offs. Trade-offs is encountered when personal or business decisions are made. In making these decisions, one has to ensure that he or she becomes better off with his or her chosen option. We also realize in our life that we cannot get all that we want and we have to make trade-offs so long as it places us in a better or preferred position. This is more of a philosophical discourse, but it goes to show that the concepts of chemical engineering can be as useful in life and in business as it is in chemical processing.

To tackle these trade-offs, the Production Possibility Frontier in economics can be used. In view of the Principle of Intrinsic Reality where growth is limited by the fixed amount of material and energy, the Production Possibility Frontier needs to shift to the right as shown in Figure 3.

Figure 3. Graphical Representation of Expanding the Production Possibility Curve

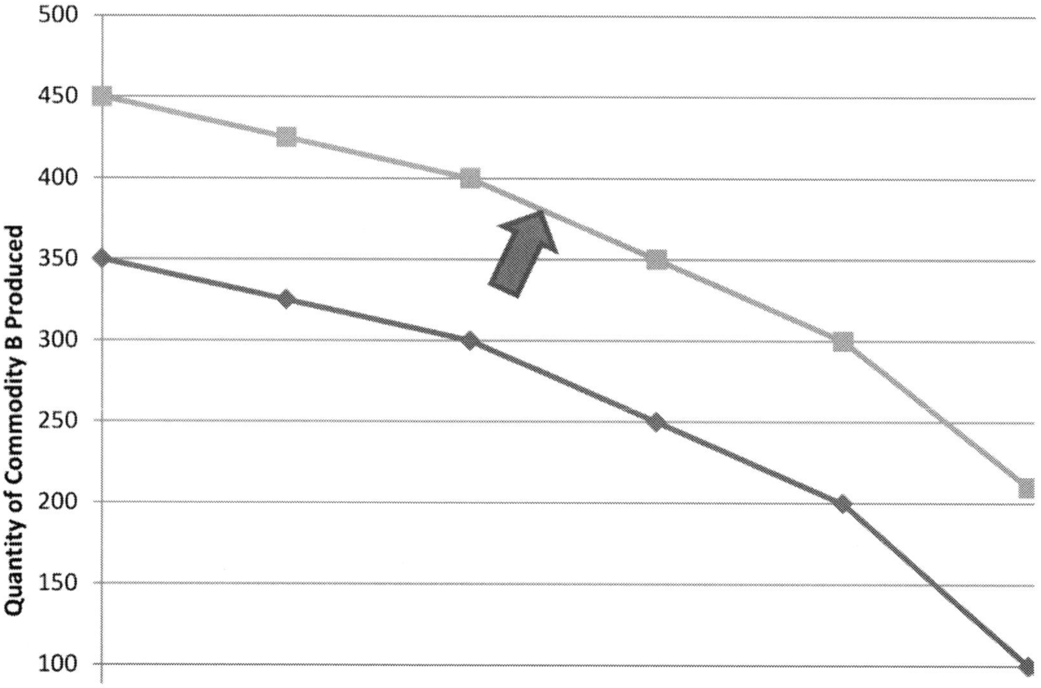

This shift has an implication on the practice of chemical engineering as Research and Development (R&D) is needed to achieve the said shift. R&D enables the economy to produce more of the two commodities (i.e. Commodity A and Commodity B) resulting in economic growth. The key challenge here to Chemical Engineering, particularly on Chemical Engineering R&D, is to look at the bigger picture where economic growth emanating from R&D output is pivoted on sustainability; considering its impact to the environment and the society at large.

3. Paradigm Shift in the Practice of Chemical Engineering

"Money" or "Profit Maximization" has been at the center of the practice of chemical engineering (Smith 2005). Various mathematical optimization methods are used to design systems, processes or components of processes with the end in mind of optimizing profits. This mental model however, will no longer work in our situation today as we face economic, ecological, social and spiritual crisis (Scharmer 2010).

To address this, Scharmer (2010) pointed out seven acupuncture points to shifting capitalism into a regenerative ecosystem economy. Within this purview, the practice of chemical engineering has a greater role to assume. For example, Scharmer (2010) suggested that the design of all production and consumption cycle be completely earth-to-earth, i.e. without necessitating for landfill and that it has to co-evolve with the natural ecosystem. In this regards, the way we do optimization should be revamped into sustainable optimization. By sustainable optimization, it means optimization of the entire system and adding in environmental as well as societal objectives to the equation.

In this definition, two things need to be considered. The first one is systems thinking which emphasizes the understanding of the entire system so the interaction among components will be known and unintended consequences mitigated (Senge 2006). This is a holistic approach to problem solving and finds enormous use in process design.

It should be noted however that the construction of processing plant, whether it is a greenfield or a brownfield project, is rooted on projected market demand. In some instances, economic volatility necessitates companies to experiment with multiple pathways to survive. In this circumstance and within the purview of systems thinking, design thinking comes into the fore. Design thinking is a collective principle emphasizing empathy for users, doing prototyping and tolerance for failure (Kolko 2015). This principle finds enormous use in chemical engineering designs such as in product development, process development, systems improvement and the like. This is also useful for simplifying or humanizing chemical engineering designs with the end in mind of contributing to sustainability. It also has to be noted that of paramount importance in using this tool is imagining the future with the view in mind of status quo and the likely outcome for sustainable development.

The second consideration is the inclusion of environmental and societal objective functions to the objective functions (economic criteria) of profit, cost, energy and yield (Edgar, et al. 2001) in optimization. The environmental objectives can be the impact or risk to environment and the societal objectives can be the impact or risk to society or common good.

These are just few examples of the needed paradigm shift in chemical engineering practice in consideration of the principle of intrinsic reality. But there can be more. A picture of the envisioned new paradigm for the practice of chemical engineering is shown in Figure 4.

Figure 4. New Paradigm in the Practice of Chemical Engineering

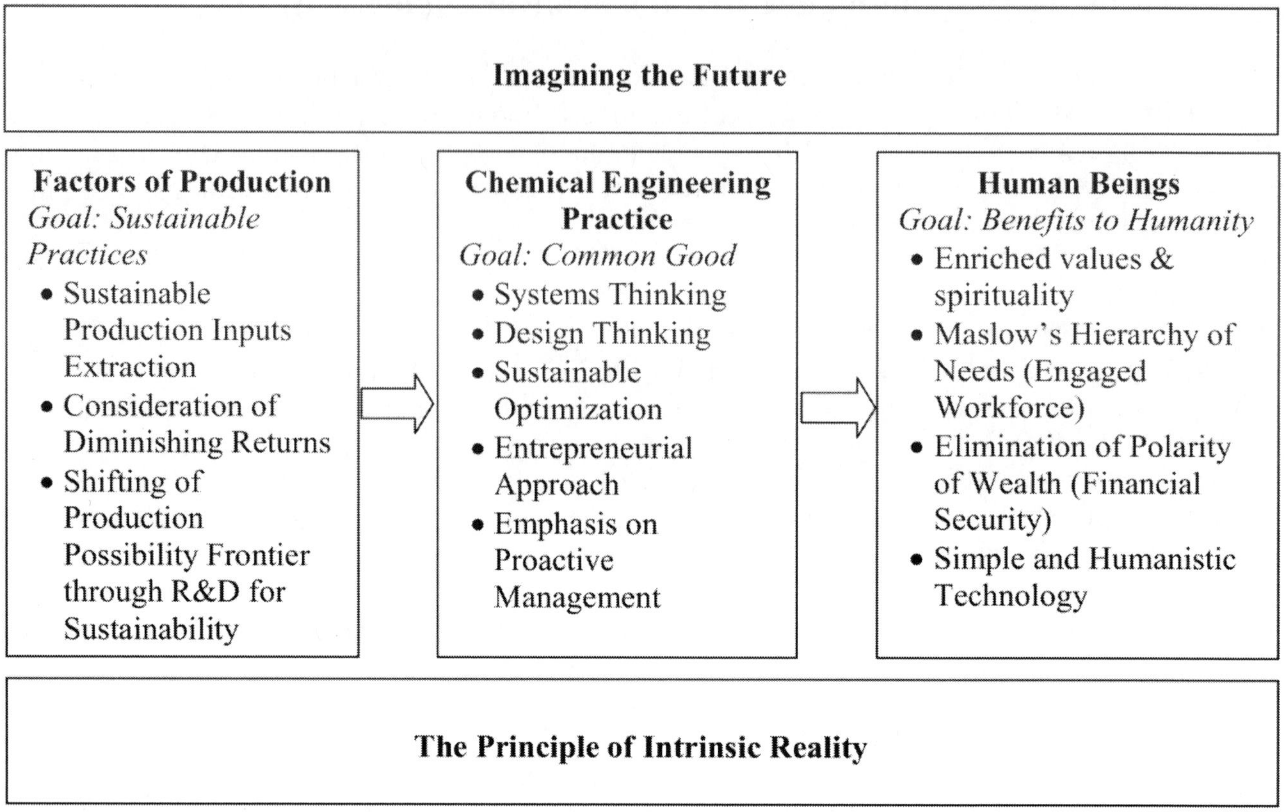

It should be realized that at the core of our economic systems is chemical engineering which can drive the economy into positive direction with the least unintended consequences. Related to that is capitalism in which upgrading of its operating systems can affect the practice of chemical engineering to a substantial degree. A discussion of this is provided in the following section.

4. Wealth Creation and Its Imperative for Sustainability

The development of this paper was inspired by the book "Freakonomics" in which the authors Steven D. Levitt and Steven J. Dubner (2005) asserted that economics can explain almost everything. It came to mind that if economics as a basic science can explain almost everything, then why not chemical engineering which was founded on chemistry, the central science?

Reflecting on it deeper, while explanations clear our doubts, we need more than that so we can tackle the world's woes. As such, this author thought that perhaps combining the two disciplines of chemical engineering and economics can be a potent force towards this end. How this can possibly be done is discussed in this section.

The starting point here is a review of the economic history of the world. In his book "A Farewell to Alms", Gregory Clark (2007) reviewed the economic history of the world in which he divided it into two periods; the period before 1800 and the period after. This is graphically represented in Figure 5.

Figure 5. Shares of Major Economies in World GDP through Time

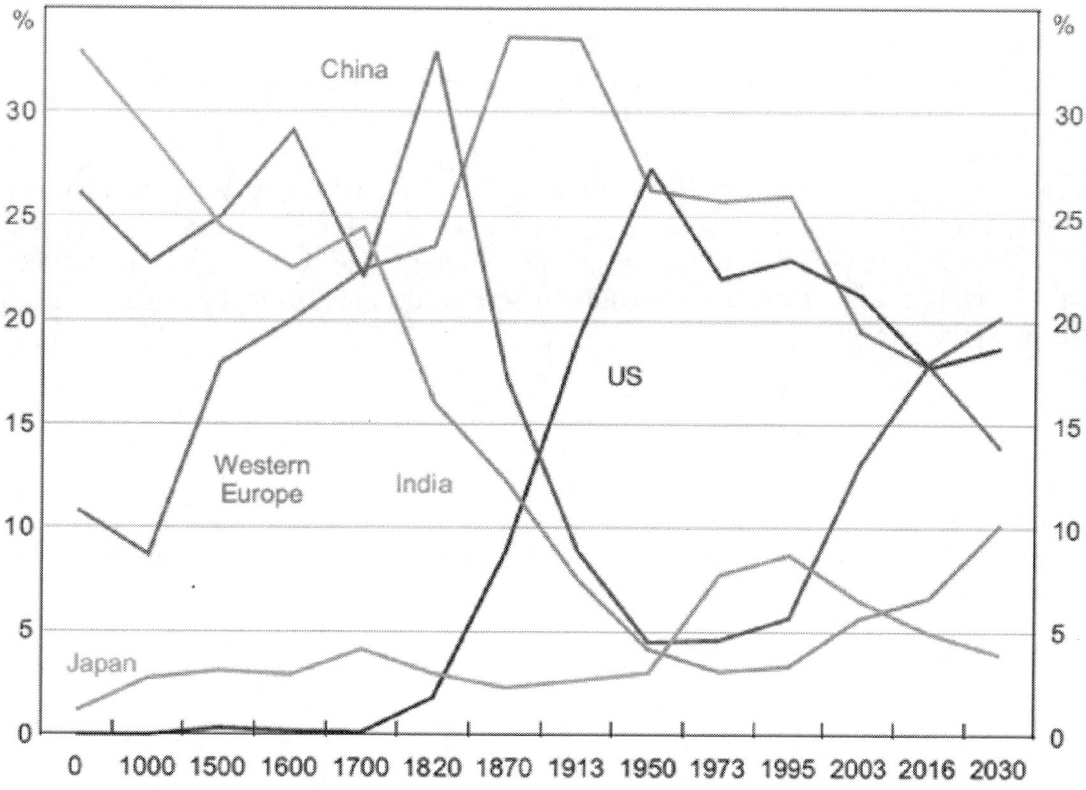

Note: Based on purchasing power parity (PPP) valuation of country GDP

Sources: International Monetary Fund (IMF), *World Economic Outlook* database, April 2011; Maddison (2006)

Thompson (2012) interpreted this graph by stating that "everything to the left of 1800 is an approximation of population distribution around the world and everything to the right of 1800 is a demonstration of productivity divergences around the world." In other words, everything to the left before Industrial Revolution depicts economic power by the sheer size of population and everything to the right after the Industrial Revolution shows the effect of mass production on country's economic output.

Clark (2007) on the other hand called the era before the industrial revolution, i.e. ca. 1800, as Malthusian era in which income per person varied across societies and epochs but growth is not observed. As Clark (2007) explained, this was due to a mechanism called Malthusian trap in which short-term increase in income through advancement in technology is dragged down by population growth. He further explicitly stated:

> "Economics as a discipline arose in the dying decades of the Malthusian era. Classical economics was a brilliantly successful description of this world. But the torrent of goods unleashed by the Industrial Revolution not only created extremes of wealth and poverty across nations, it also undermined the ability of economic theory to explain these differences."

In view of this statement, Figure 5 is reinterpreted in this paper. The contention of this author is that *polarity of wealth* can be gleaned from the graph. *Polarity of wealth* means the uneven concentration of wealth in the global economic system as some parts of the world are richer than others. The polarity can also be observed to shift as some richer parts of the world decline in economic power and some poorer parts of the world rise in economic wealth.

A historical example can be gleaned from Figure 5. India and China up until 1500 CE had 50 and 60 percent share of the world's economy, respectively. In 1 CE, China had a population of almost 60 million, while the area of the present day United States had a population of 680,000. In the late 18th century, the Industrial Revolution rendered countries with large population, just countries with the largest populations.

Looking at North America, it took them 1800 years to overtake China's economic output. However, by 1950, the US economic output was three times as great as China, even though its population is three times less. Comparing China with UK in 1913, the UK had a population of 45 million and China had 437 million but the economic output was almost the same. Analyzing what happened in Europe and later in the United States, Europe was on the lead even before 1800 (Nye 2012).

In 2010 according to data, China overtook Japan as the second biggest economy, though the numbers may not be strictly comparable. Japan's data are seasonally adjusted while those for China were not. Angus Maddison suggested that China and India were the biggest economies in the world for most of the past 2000 years. That is why economists find it a mystery why they fell so far behind and now they are flourishing ('China's economy' 2010). This may not be mysterious considering the *polarity of wealth* as was mentioned earlier and that this polarity can be argued to emanate from the Principle of Intrinsic Reality.

The reasoning here is that the world has constant mass and energy. Assuming that all these mass and energy can be harnessed to create wealth, it could mean that there is a limit for wealth creation. Theoretically, the limit can be expanded indefinitely through R&D but experience demonstrates that this was not achieved yet or getting anywhere near being achieved. As such, the constant mass and energy vis-à-vis the limit for wealth creation can be thought of as the reason for the polarity of wealth, i.e. wealth can only shift from one entity to another.

The implication of this assertion can be demonstrated by the fact that the 1% richest have accumulated more wealth than the rest of the world combined (Oxfam International 2016). The world has never seen 100% of its population to be wealthy. As such, time and again, equitable distribution of wealth is called for and mechanisms to achieve such_should be made a priority in government policies.

In a more concrete way, Scharmer (2010) suggested measures to address this growing inequality which is consistent with the mindset espoused by the Principle of Intrinsic Reality. These measures are as follows:

- Creation of a New Class of Common Property Rights

The creation of common property rights is intended to address the cost to society of private economic undertakings. For example, a private company is extracting natural resources from a certain place. While the private company can profit from that endeavor, the community in that place might suffer the undesirable consequences (e.g. flooding due to forest denudation). To ensure sustainability and institutionalize the voice of future generations in this case, a trust can be established which will own the rights and distribute funds (e.g. from cap-and-trade system for carbon emissions) to citizens (one share per person), communities and the government for investment in public infrastructure.

The trust is an effective means because it does not function like a company (driven by profits) or government (driven by special interest groups) but that it is accountable for long-term sustainability of the specific commons it manage for future generations.

- Creation of a Human Right to Basic Income

One way to ensure equitable distribution of wealth is to create a human right to basic income for every human being. This right when combined with free or affordable access to healthcare and education would create a playing field where everyone has a chance to pursue aspirations and dreams. In other words, it enables individuals to put their real creativity for the benefit of larger community. Another intention of this measure is to free up paid labor (a form of dependence) so that human beings can enhance their creativity and entrepreneurial skills.

- Inclusive Capitalism

To eradicate the *polarity of wealth*, capitalism should be made inclusive so that money and capital flows are directed to all sectors of the economic system. Currently, our money system leads to accumulation of oversupply of money and capital in high financial but low environmental and social gain areas. At the same time, there is an undersupply of money and capital in areas that address societal and community needs. It implies that there is too much money in areas where it is not needed but not enough on areas where it is needed. To address this imbalance, it is recommended to redesign the system so that the current extraction-based financial system is replaced by institutions which can support the real economy and thereby serve the ecology and community.

It should be noted that shifting of wealth is also observed in various geographical regions. When this occurs, geo-political instability can result. This goes to show that if we want lasting stability and security within the limits of our intrinsic reality, then we should not tolerate unabated greed. Otherwise, *polarity of wealth* will continue to persist and we cannot realize our full potential as individuals and collaborators for common good.

5. Way Forward...

Recognizing our intrinsic reality leads us to greater understanding of our very nature bounded by the constancy of mass and energy. It shed light on why growth stabilizes after addition of more inputs. Within this purview, trade-offs are also encountered which necessitates for R&D. R&D initiatives however should be underpinned on sustainability.

Looking at the practice of chemical engineering from the lens of intrinsic reality, the approach should be futuristic while considering relevant aspects related to factors of production and human beings.

The bigger picture shows that Engineers particularly Chemical Engineers are major players in driving the real economy. As Scharmer (2010) aptly puts it, "The essence of technology and knowledge creation, and the ultimate driving force of the economy, is the same: it is the human capacity to create." As such, if sustainability is to be achieved, Chemical Engineers have a greater responsibility to innovate for regenerative and social technologies. In this way, we can stem the polarity of wealth by providing affordable technology to enable disadvantaged individuals enhance their creativity, innovation and entrepreneurship. This can help them partake in the benefits of inclusive capitalism.

As demonstrated in this paper, the synergy emanating from the convergence of Chemical Engineering and Economics can powerfully tackle the world's woes. As such, there is a good reason for chemical engineers and economists alike to collaborate particularly on sustainability initiatives.

6. References

[1] 'China's economy overtakes Japan's in real terms' 2010, *The Economist*, 16 August, viewed 5 March 2016, http://www.economist.com/node/16834943

[2] Clark, G 2007, *A Farewell to Alms: A Brief Economic History of the World,* Princeton University Press, Princeton, NJ.

[3] Edgar, T, Himmelblau D & Lasdon, L 2001, *Optimization of Chemical Processes*, McGraw-Hill, New York.

[4] Huang, Y & Wang, B 2011, 'From the Asian Miracle to an Asian Century? Economic Transformation in the 2000s and Prospects for the 2010s', in H Gerard & J Kearns (eds), *Proceedings of the Reserve Bank of Australia annual conference*, Reserve Bank of Australia, Sydney, viewed 30 January 2016,

http://www.rba.gov.au/publications/confs/2011/huang-wang.html

[5] Kolko, J 2015, 'Design Thinking Comes of Age', *Harvard Business Review*, pp. 66-71, viewed 24 February 2016, https://hbr.org/2015/09/design-thinking-comes-of-age

[6] K.N.C. 2012, 'Mis-charting economic history: more 2,000 years in a single graphic', *The Economist blog*, web log post, 20 June, viewed 29 February 2016,

http://www.economist.com/blogs/graphicdetail/2012/06/mis-charting-economic-history

[7] Levitt, S & Dubner, S 2005, *Freakonomics: A Rogue Economist Explores the Hidden Side of Everything*, William Morrow, New York, NY.

[8] Malthus, T 1798, *An Essay on the Principle of Population*, J. Johnson in St. Paul's Church-yard, London.

[9] Maslow, A 1943, 'A Theory of Human Motivation', *Psychological Review*, vol. 50, pp. 370-396, viewed 24 February 2016, http://psychclassics.yorku.ca/Maslow/motivation.htm

[10] Monod, J 1949, 'The Growth of Bacterial Cultures', *Annual Review of Microbiology*, vol. 3, pp. 371-394, viewed 2 February 2016,

http://www.annualreviews.org/doi/abs/10.1146/annurev.mi.03.100149.002103

[11] Nye, J 2012, 'Fascinating new graph shows the 'economic history of the world since Jesus'', *MailOnline*, 23 June, viewed 15 November 2015,

http://www.dailymail.co.uk/news/article-2163610/Fascinating-new-graph-shows-economic-history-world-Jesus.html

[12] Oxfam International 2016, *An economy for the 1%, how privilege and power in the economy drive extreme inequality and how this can be stopped*, Oxford, viewed 5 March 2016,

https://www.oxfam.org/sites/www.oxfam.org/files/file_attachments/bp210-economy-one-percent-tax-havens-180116-en_0.pdf

[13] Samuelson, P & Nordhaus W 2009, *Economics*, McGraw-Hill Higher Education, New York, NY.

[14] Scharmer, O 2010, 'Seven Acupuncture Points for Shifting Capitalism to Create a Regenerative Ecosystem Economy', *Oxford Leadership Journal*, vol. 1, pp. 1-21, viewed 24 February 2016, http://www.oxfordleadership.com/journal/vol1_issue3/scharmer.pdf

[15] Senge, P 2006, *The Fifth Discipline: The Art & Practice of the Learning Organization*, Doubleday/Currency, New York.

[16] Smith, R 2005, *Chemical Process Design and Integration*, John Wiley & Sons Ltd., West Sussex, England.

[17] Thompson, D 2012, 'The economic history of the last 2,000 years in 1 little graph', *The Atlantic blog*, web log post, 19 June, viewed 29 February 2016, http://www.theatlantic.com/business/archive/2012/06/the-economic-history-of-the-last-2-000-years-in-1-little-graph/258676/

Acknowledgment

The author would like to thank Hyundai Engineering and Construction Co. Ltd. for the support in this endeavor. The author is also thankful to Ms. Sofia Ayu for meticulously editing this manuscript and Mr. Marcial N. Barroga for the poster design.

445030 Heavy Metal Detection in Treated Water Via a Fabric Nanocomposite Sensor

Abstract:

We present the conception of an electrochemical sensor for on-site/portable water analysis. This flexible sensor was developed to quantify the amount of heavy metals (in real-time), alleviating the wait time, large sample size, possible contamination, and expensive analytical equipment associated with current procedures for water analysis. This sensor is composed of a multiwalled carbon nanotube (MWNT) functionalized polymer mat, for a flexible yet conductive sensor. The MWNTs are then functionalized with a cyclo-oligomeric calixarene, which is shown to selectively form a supramolecular complex with metal ions. Upon complex formation, the charge carriers are drawn away from the carbon layer and the detection is prevalent at appropriate levels. The optimization of MWNT and calixarene functionalization, as well as the selectivity and sensitivity of the sensor will be discussed.

444086 Optimization of Layered Graphene Oxide Nanostructures through Controlling the Membrane Deposition Rate Towards Size-Sieving Separation

Abstract:

Single-layer graphene oxide (GO) flakes were coated on anodized aluminum oxide (AAO) substrate through a simple vacuum filtration process, the effect of the deposition rate on the nanostructures of obtained membrane was investigated though pure water flux and organic vapor diffusion, which indicated GO membrane assembled with slow deposition rate might have more ordered lamellar structure and, correspondingly, have smaller effective pore size. In contrast, under fast deposition rate, GO flakes assembled in a more random way. Therefore, through controlling the membrane deposition rate, the nanostructure of layered GO membrane could be optimized for size dependent separation. We found after thermal reduction, 100 nm GO membrane (prepared @ 0.02 nm/s) has effective pore size smaller than 6.1 Å, which exhibit great potential in n-hexane/2,2-dimethybutane (DMB) separation.

444376 Ozone Source Apportionments of Local Anthropogenic Emissions

Abstract:

Ozone is one of six principal air pollutants designed by US EPA. NOX and VOC emissions from anthropogenic sources are the most important factors to control ozone pollutant.

In this report, photochemical dispersion models are used to estimate the effect of ozone from local anthropogenic emission sources. EKMA-type diagrams are generated for NOX- and VOC- effects. Furthermore, APCA (Anthropogenic Precursor Culpability Assessment) technology is used to estimate ozone source apportionments.

EKMA-type diagrams show that 10-30% emission reductions could reduce ozone concentrations to meet standard. Further APCA results show the average ozone contributions from regions and emission categories. APCA results explain different effects for different monitoring stations.

444142 Robust Design of a Radiant Wall Burner for a High Temperature Environment Utilizing Computational Fluid Dynamics Simulations

Abstract:

Radiant Wall Burners are often utilized in Ethylene Cracking Furnaces. The conditions and fuel gases for a specific unit can vary considerably within the unit and across similar units. With burner air preheated to high temperatures, premixing the fuel can result in dangerous and destructive flashback, while complete nozzle-mixing of the fuel can result in burner instability at startup or turn-down conditions. These challenges can be solved through the generation of a robust radiant wall burner design, which is facilitated by executing Computational Fluid Dynamics simulations of many different geometric configurations.

444397 Flare Minimization and Emission Characterization during Start-Ups of Chilling Train in Ethylene Plants

Abstract:

Flaring is a crucial safety operation to protect chemical plant personnel and equipment during plant start-up. However, it generates huge amounts of off-spec products, and causes negative environmental and social impacts, as well as results in tremendous raw material and energy losses which could be saved to generate much more needed products. Thus, cost-effective start-up flare minimization strategies which is a double-win practice, simultaneously benefiting both the industrial and environmental sustainability, are becoming more and more important and attractive to the industries.

The chilling train is a very important section for the ethylene plants and accounts for more than 50% of the total start-up time and flare emissions of the whole ethylene plants. Within this section, the charge gas from the upper stream process is refrigerated to a very low temperature (lower than -160 °C) to separate hydrogen and methane. Meanwhile, the left charge gas is liquidized for cryogenic separation. The chilling train system consumes over 40% of the refrigerating capacity of an ethylene plant. During the chilling train start-up, different operational conditions with different compositions and flowrate of precooling process will affect both product separation efficiencies, start-up stability time and flare emission.

Traditional start-up scenarios are developed based on experience and improved by trial and error, which is inefficient and dangerous. In this study, a general methodology on flare minimization for chilling train start-up operations via rigorous dynamic simulation is presented. The methodology starts with setup and validation of chilling train steady-state and dynamic simulation models. The initial state of start-up in the real plant is achieved through systematically transforming from the validated dynamic model, and several operation conditions with different compositions and flowrate of precooling stream are examined to identify an optimal initial state of start-up based on the obtained initial state. Thereafter different flare minimization strategies for the chilling train start-up are conducted on the optimal initial state model. Also, emission source characterization under different flare minimization strategies for chilling train start-up is investigated. Deep insights of the start-up strategies, emission source distribution and dynamic emission profiles are provided. This study provides detailed technical support for both the industry and environmental agencies on evaluating and developing cost-effective flare minimization strategies in the future.

439811 Purification of Ibuprofen Enantiomers By a New Technique Combining Distillation and Crystallization

Abstract:

Stripping crystallization (SC) is a new separation technique which combines distillation and crystallization. Basically, SC is operated at a triple-point condition, in which the liquid is simultaneously vaporized and crystallized due to the three-phase equilibrium. In essence, the process is continued until the liquid phase is completely eliminated and only the pure solid crystals remain in the feed. Thus, crystal washing is not required since only pure solid crystals remain in the feed and no impurities are adhered on the crystal surfaces at the end of SC. In the previous research, SC has been successfully applied to separate p-xylene from the mixed xylenes by Shiau et al. (2005, 2008, 2013).

Ibuprofen, also called 2-(isobutylphenyl)-propionic acid, is a non-steroidal anti-inflammatory drug widely used to treat headaches and minor pains. Although it is known that S-ibuprofen is responsible for the anti-inflammatory effects while R-ibuprofen might cause some long-term side effects, ibuprofen is still sold as a racemic mixture in the market due to the complexity involved in the separation of enantiomers. A thermodynamic model is developed in this work to simulate the three-phase equilibrium conditions during the SC operation. The experiments show that seeding with ultrasound mixing or magnetic stirring can be efficiently employed to purify S-ibuprofen from ibuprofen enantiomers. The experimental results, including the final enantiomeric purity and recovery ratio of S-ibuprofen, are consistent with the simulation results predicted by the model.

References:

1. Shiau, L. D.*, Wen, C. C. and Lin, B. S. "Separation and purification of p-xylene from the mixture of m-xylene and p-xylene by distillative freezing", *Ind. Eng. Chem. Res*, **44(7)**, p2258-2265, 2005.

2. Shiau, L. D.*, Wen, C. C. and Lin, B. S. "Separation of p-xylene from the multicomponent xylene system by stripping crystallization", *AIChE Journal*, **54(1)**, p337-342, 2008.

3. Shiau, L. D.*; Liu, K. F. "Investigations into the effects of the cooling rate on stripping crystallization", *Ind. Eng. Chem. Res*, **52**, p1716-1722, 2013.

442557 Flow Field Simulation and Distribution Performance Analysis of a Gas Pipe Distributor in Fccu

Abstract:
A fluidized catalytic cracking (FCC) is an important petroleum refinery process, which, aided by highly selective zeolite catalysts, cracks large-molecule low-value heavy oil feedstock into small-molecule highvalue
products. When spend catalyst is fed into the regenerator for reactivating, it is fluidized by the gas distributor at the bottom of regenerator. The gas distributor is designed to uniformly distribute gas to maintain the fluidized state of the fluidized bed and to provide the reaction gas to guarantee the burning coke deposited on the spend catalyst. Therefore, the gas distribution performance of the gas distributor influences the quality of initial gas distribution directly. So the pipe distributor used in a industrial FCC unit, which is composed by main pipes and a number of branch pipes with many nozzle tubes, was investigated by means of flow field calculation. The gas distribution performance of the pipe distributor was analyzed based on the simulated flow field. The simulation result shows that there are a obvious nununiform
gas distribution from the different nozzle gas flow rate. The static pressure increases gradually along the flow direction in the nozzle tube while the axial flow rate decreases, which leads to the nozzle gas flow rate increasing gradually. Furthermore, an obvious bias flow occurs at the inlet of the nozzle tube, which influences the gas flowing into the nozzle. The pressure drops performance of nozzle and pressure varying in the branch pipe are analyzed comprehensively. Due to the nonuniform
gas distribution by the gas pipe distributor in the practical usage, the problems such as overlarge temperature difference among the fluidized bed layers in the regenerator have often happened, which will not only reduce burning coke efficiency and but also decrease activity of the regenerated catalyst. Meanwhile, the erosion of the distributor also occurs, and in consequence the function of the distributor fails, which even causes the nonscheduled shutdown of the FCC unit and leads to tremendous economic loss. The support is provided for optimizations of both structural parameters and operation parameters of the gas pipe distributor and improvement of the gas distribution performance. Furthermore, improvement suggestions were provided based on the above reasoning. The results are expected to be useful for the distributor design.

441535 Hydrogen Disposal Flare System Hazard Identification and Mitigation Analysis

Abstract:

A hazard assessment on a hydrogen disposal flare system was conducted in preparation for a hydrogen turbopump test at the Air Force Research Laboratory (AFRL), Aerospace Systems Directorate at Edwards AFB, CA. The test operations required the disposal of large quantities of liquid and gaseous hydrogen through a commercial flare system to prevent the formation of an explosive environment. As part of the assessment, the primary steps in the test operation process were evaluated for the formation of a potential hazard by looking at each associated failure mode and the resulting affect on personnel and facilities. Evaluation of each operational step was conducted in a Process Failure Modes and Effects Analysis (PFMEA) that identified the probability and severity of the identified effects and associated hazards. If the probability and severity were unacceptable, mitigating steps were applied until the overall hazard risks were reduced to acceptable levels. The recommendations of this assessment were incorporated into both the flare system design and the operational procedures. This paper outlines the results of the PFMEA which include identification of the effects/hazards, available risk-mitigating techniques, and implementation of these techniques used for the mitigation of the identified hazards. Based upon the findings significant design modification to the vender's commercial equipment was needed to comply with recent changes in the code. Most of the identified hydrogen flare system hazards/effects were categorized into four groups: internal explosions, obstructions, external explosions, and maintenance degradation. Many corrective actions were implemented to mitigate both the probability and severity of the identified hazards. In addition, this paper discusses the installation of oxygen monitoring systems on the flare system, determination of appropriate system purge requirements, installation of purge system status alarms, incorporation of a triple redundant flare ignition system with status indicators, and a number of other hazard risk mitigation techniques. This comprehensive description of the hazards and mitigation of a hydrogen disposal system is a plant is vital part to safe operation of both hydrogen commercial and aerospace facilities.
<align="center">

444020 Thermodynamic Modeling of Aqueous Lithium Sulfate and Its Solubility in the Presence of Na+, K+ and Mg2+ Sulfates

Abstract:

This work is an extension of the previous works done by our group towards development of a rigorous and comprehensive thermodynamic model for high salinity produced water that is essential for mass and heat balance calculations and simulation and design of novel treatment processes. Specifically this work presents a thermodynamic model that accurately represents all thermodynamic properties of aqueous lithium sulfate solution including the solubility of Li_2SO_4 in water and also in the presence of Na^+–K^+–Mg^{2+} sulfates. By explicit consideration of solution chemistry of each subsystem, symmetric electrolyte NRTL model is used to correlate available experimental data of all pertinent thermodynamic properties including activity coefficient, osmotic coefficient, excess enthalpy, and liquid heat capacity over the entire concentration range. This facilitates quantification of eNRTL binary molecule–molecule, molecule–electrolyte, and electrolyte–electrolyte interaction parameters. Also, solubility data are used to regress the thermodynamic constants required for modeling the liquid–solid equilibrium of lithium sulfate in water and in the presence of Na^+–K^+–Mg^{2+} sulfates. Displaying a significant prediction ability, the model accurately represents the thermodynamic behavior of the system from infinite dilution up to saturation and temperatures from 298.15 K to 573.15 K.

442068 Enhanced Furfural Production By an Innovative Bottom Dividing Wall Column with a Decanter Configuration

Abstract:

Enhanced Furfural Production by an Innovative Bottom Dividing Wall Column with a Decanter Configuration

Nguyen Van Duc Long, Le Cao Nhien, Le Quang Minh and Moonyong Lee†

School of Chemical Engineering, Yeungnam University, Gyeongsan 712-749, South Korea

Furfural, which are used to derive other industrial chemicals, has been identified as one of the major bio-based platform chemicals that can compete with petroleum-based chemicals. However, the current commercial furfural process has low yield and requires intensive energy. This paper reports the simulation, detailed design and optimization of an industrial process recovering furfural from lignocellulosic biomass, a renewable and inexpensive feedstock. The aqueous feed mixture from the acidic hydrolysis of biomass was fed to the separation process to recover furfural as the main product, along with methanol as valuable byproducts. The purification process was intensified by applying an innovative bottom dividing wall column with a decanter configuration (BDWC-D). The response surface methodology (RSM), which allows the interactions between variables to be identified and quantified, was used to optimize the BDWC-D structure. The predictions by the RSM showed satisfactory agreement with the rigorous simulation results. The heat pump technology and heat integration were then employed to improve the purification process. The results showed that the proposed sequence could save the overall energy requirements and total annual cost significantly as compared to the conventional sequence. This sequence can be employed for both grass-root and retrofit designs.

Keywords: Lignocellulosic Biomass; Furfural; Energy Efficiency Improvement; Dividing Wall Column; Response Surface Methodology

Acknowledgments

The authors wish to acknowledge the financial support from the R&D Convergence Program of NST (National Research Council of Science & Technology) of Republic of Korea and KITECH (Korea Institute of Industrial Technology) (ES150001).

*** Formatting Issue - Best Available Paper/Graphic ***

444278 Insight into the Mechanism of Toxic Chromium (VI) from Wastewater Using Magnetite Coated PINE Powder

Abstract:

INSIGHT INTO THE MECHANISM OF TOXIC CHROMIUM (VI) FROM WASTEWATER USING MAGNETITE COATED PINE POWDER

Pholosi A., Ofomaja A., Naidoo E.B.

Biosorption and Wastewater Treatment Research Laboratory, Department of Chemistry,

Faculty of Applied and Computer Sciences, Vaal University of Technology, P. Bag X021, Vanderbiljpark, 1900 South Africa.

INTRODUCTION

A knowledge of the interaction between pollutant molecules or ions and active sites on an adsorbent surface is an important phenomenon in many industrial processes such as catalysis, adsorption and other separation techniques. The use of lignocellulosic materials for Cr(VI) adsorption has gained interest over the years **(Miretzky and Cirelli, 2010)**. The adsorption properties of lignocellulosic materials has been attributed to acidic functional groups on their surfaces which can take part in ion-exchange, complexation, chelation and hydrogen bonding with pollutant species in solution **(Vithanage et al., 2015)**. Two mechanisms for lignocellulosic adsorption of Cr(VI) have been proposed; (i) an electrostatic mechanism in which at low pH, the protonated acidic functional groups on the biomaterial carrying a positive charge attracts the negatively charged Cr(VI) species **(Demirbas, 2005)** and (ii) the adsorption-coupled reduction mechanism in which Cr(VI) is reduced to Cr(III) either in solution (Direct reduction) or after being adsorbed (Indirect reduction) by electron-donor groups of the biomaterial that have lower reduction potential values than that of Cr(VI) **(Gao et al., 2008)**. The Cr(III) ions formed in solution may then form complexes with Cr(III)-binding groups on the biomaterial surface or released into bulk solution by repulsion between Cr(III) and other positively charged groups on the biomaterial **(Miretzky and Cirelli, 2010)**.

METHODS

Sodium hydroxide treated pine coated with magnetite (NTP-NC) was prepared by co-precipitation method. The effect of solution pH was performed in a batch system in which 0.5 g of the adsorbent was contacted in six 250 cm^3 beakers containing 75 cm^3 of 75 mg/dm^3 Cr(VI) solutions set at pH 1, 2, 4, 6, 8 and 10 at 26 °C and agitated at 200 rpm for 2 h and the clear solution analysed for Cr(V/I) left using UV-Vis spectrophotometry at 460 nm

RESULTS

The results of the effect of initial solution pH on the adsorption of Cr(VI) onto NTP-NC is shown in Fig. 1. To confirm the operating mechanism, the change in $[H^+]$ concentration at each initial pH, the oxidation/reduction potential before and after adsorption and the amounts of Total Cr, Cr(VI)and Cr(III) present in solution at each pH was monitored for NTP-NC and the results displayed in Fig1 . The change in $[H^+]$ concentration ($\Delta[H^+] = [H^+]_{final} - [H^+]_{initial}$) at the end of the adsorption process for each initial solution pH was recorded and plotted against the initial solution pH. The results showed that the change in $[H^+]$ concentrations at the end of the adsorption process where higher for lower initial solution pH's (where lower amounts of Cr was found in solution) than for higher initial solution pH's (where higher amounts of Cr was found in solution) and the change in $[H^+]$ concentration became small and almost constant after initial solution pH 4. This is consistent with the assumption that the removal of Cr from solution is associated with [H^+] ion consumption and that as the concentration of [H^+] ions reduces in solution, the Cr removal reduces. Cr(VI) ion in acidic solution has a positive redox potential value (above 1.3 V at standard conditions) and in the presence of an electron donor becomes unstable and can be reduced to Cr(III) **(Park et al., 2008)**. The results show that the initial ORP values (before adsorption) for the Cr solutions at pH's of 1, 2, 4, 6, 8, and 10 were 339, 280, 165, 55, -53 and -172 mV for NTP-NC respectively. The positive values of ORP at lower solution pH's indicates higher possibility for reducing Cr(VI), while the lower and negative ORP values indicates lower possibility for reducing Cr(VI). On addition of the adsorbents into the solutions, a change in ORP were observed . The values of ORP at the end of the adsorption process became 330, 268, 30, 9 -29 and -61 for NTP-NC respectively. A reduction in ORP at the end of the adsorption stage signifies that reduction has occurred in the system. **Yu et al. (2014)** showed the amount of Cr(VI) removed from aqueous solution in contact with zero-valent iron (nZVI) could be related to the change in ORP and pH of the solution. The authors observed a decrease in ORP from 400 to 100 mV and a corresponding 10 unit pH increase as Cr(VI) was removed from solution.

Finally, analysis of the prepared Cr solutions at the various solution pH's showed that the concentration of Total Cr obtained from ICP and that of Cr(VI) using UV-vis measurements were very similar, suggesting that Total Cr was almost equal to Cr(VI) in solution. After contacting the various solutions with the adsorbents, the concentration of Total Cr in the pH 1 solution was observed to have reduced to 21.56 mg/dm^3 for NTP-NC, while the Cr(VI) concentrations was 10.66 mg/dm^3 for NTP-NC. With solution pH 2, the concentration Cr(III) left in solution was lower than that of solution pH 1 for both samples. The reason is that at pH 1 there is an abundance of H^+ ions in solution which accumulates on the functional groups on the adsorbent making the surfaces of the adsorbent positively charged and repelling away the positively charged Cr(III) ions from the adsorbent surfaces **(Suksabye et al., 2009)**. At solution pH 2, it is believed that there is less accumulation of H^+ ions at the adsorbent surfaces which led to reduced electrostatic repulsion and increased adsorption of Cr(III) from solution. This increase in Cr(III) adsorption is also confirmed by the reduction in Total Cr in solution between solution pH 1 and 2. The concentrations of Cr(VI) left in solution at pH 2 was higher than at pH 1 for NTP while for NTP-NC the Cr(VI) concentrations were lower at pH 2. For solution pH's higher than 2 and up to 10, the residual Cr(III) in solution becomes very low and almost constant. This is consistent with the small change in [H^+] in solution which is almost constant at this pH range.

FTIR Evidence for the adsorption of Cr(III)

When Cr(III) is formed in solution via adsorption-coupled reduction mechanism it may undergo hydrolysis and/or complexation depending on the solution pH and the total Cr(III) concentration. Hydrolysis may lead to the formation of various Cr(III) species with an increase in solution pH which gives rise to a reduction in its solubility and increasing the chances of precipitation of the Cr(III) species. At solution pH 1 – 4, Cr^{3+} and $CrOH^{2+}$ are the major species which can be adsorbed by negatively charged ligands on the biomaterial surface. On the other hand, at solution pH range of 4 – 10, Cr(III) species (hydroxo complex), such as $CrOH^{2+}$, and $Cr(OH)_3$, are the dominant forms in the aqueous system and at high pH is readily transformed into soluble complex, . When NTP-NC was contacted with Cr(VI) solution for 2 hr, the final solution pH was increased to pH 2.53. The FTIR spectra for NTP-NC (Fig. 2) shows the presence of characteristic Fe-O peaks of bulk Fe_3O_4 at 553.26 cm^{-1}, 564.36 cm^{-1}, 570.21 cm^{-1} and 576 21 cm^{-1}. Functional groups associated with the pine includes the broad band at 3250.99 cm^{-1} representing –OH of cellulose, 2905.17 cm^{-1} representing C-H stretching, while the peaks at 1627.69 and 1602.87 cm^{-1} shows the COO^- and C=O bonds respectively. After 2 hr contact with Cr(VI) solution, the peak at 564.34 cm^{-1} was observed to have split into several new peaks (Fig. 3b (ii)). These emerging peaks are characteristic of Fe-O bonds of maghemite (γ-Fe_3O_4), for example the characteristics peaks of maghemite was observed at 595.21, 685.65 cm^{-1} and 582.70, 634.91 cm^{-1} suggesting a change in the iron oxide phase (Fe_3O_4 to γ-Fe_2O_3) due to the conversion of Cr(VI) to Cr(III) by magnetite nanoparticles. According to **Peterson et al. (1997)**, Cr(III)(hydr)oxide produced is precipitated on the magnetite surface. In this study, α-Cr_2O_3 peak were observed at 604.63 cm^{-1} indicating that Cr(III) was formed during the adsorption process. Finally it was also observed that the carboxylate group at 1635.55 cm^{-1} was shifted to 1637.73 cm^{-1} while a new peak was observed at 1737.72 cm^{-1} suggesting that complexation between Cr(III) and Carboxylate ions also occurred.

REFERENCE

Demirbas, A. Energy Source 27 (2005) 1449–1455.

Miretzky, P., Cerelli, A.F. A review. J. Hard. Mater. 180 (2010) 1-9.

Vithanage, M., Rajapaksha, A.U., Ahmad, M., Uchimiya, M., Dou, X., Alessi, D.S., Ok, Y.S. J. Environ. Manage. 151 (2015) 443-449.

D. Park, C.K. Ahn, Y.M. Kim, Y.-S. Yun, J.M. Park, J. Hazard. Mater. 160 (2008) 422–427.

P. Suksabye, A. Nakajima, P. Thiravetyan, Y. Baba, W. Nakbanpote.. J. Hazar. Mater. 161 (2009) 1103–1108.

R.F., Yu, F.H., Chi, W.P., Cheng, J.C. Chem. Eng. J. 255 (2014) 568–576

M.L. Peterson, A.F. White, G.E. Brown, G.A. Parks Environ. Sci. Technol. 31 (1997) 1573–1576.

Fig. 1 Effect of solution pH on the adsorption of Cr(VI)

*** Formatting Issue - Best Available Paper/Graphic ***

Fig, 2: FTIR of NTP-NC

443101 Thermodynamic Modeling of HNO_3-H_2so_4-H_2o Ternary System with Symmetric Electrolyte NRTL Model

Abstract:

Nitric acid concentration/sulfuric acid concentration (NAC/SAC) process has been widely used for concentrating dilute aqueous nitric acid and recovering spent sulfuric acid. Nitric acid is concentrated by employing sulfuric acid to bind water and break nitric acid-water azeotrope at approximately 68 wt. % nitric acid. To support heat and mass balance calculations and process simulation for NAC/SAC processes, we develop a comprehensive thermodynamic model for nitric acid-sulfuric acid-water ternary system based on previously published thermodynamic models of nitric acid-water [Wang et al., Fluid Phase Equilibria, (2015) (DOI: 10.1016/j.fluid.2015.04.015)] binary and sulfuric acid-water [Que et al., Journal of Chemical & Engineering Data, 56 (2011) 963-9772] binary systems developed with symmetric electrolyte NRTL (eNRTL) activity coefficient model. The present work takes into account one-step hydration of nitric acid molecules, two-step hydration of sulfuric acid molecules, and further hydration of hydronium ions. With up to three temperature coefficients for each eNRTL binary interaction parameter and up to two temperature coefficients for each chemical equilibrium constant, the ternary model is validated with boiling point data from 383.15 to 471.45 K and activities at 273 K. Furthermore, contour plots of boiling points, vapor phase composition, and specific heat capacity of the ternary system are generated for engineering use.

Keywords:

Nitric acid, Sulfuric acid, Electrolyte NRTL model, Vapor-liquid equilibrium, Boiling point, Heat capacity, Nitric acid concentration/sulfuric acid concentration (NAC/SAC)

444551 Continuously Adjustable, Molecular-Sieving "Gate" on 5A Zeolite for Distinguishing Small Organic Molecules By Size

Abstract:

Zeolites/molecular sieves with uniform, molecular-sized pores are important for many adsorption-based separation processes. Pore size gaps, however, exist in the current zeolite family. This leads to a great challenge of separating molecules with size differences at ~0.01 nm level. Here, we report a novel concept, pore misalignment, to form a continuously adjustable, molecular-sieving "gate" at the 5A zeolite pore entrance without sacrificing the internal capacity. Misalignment of the micropores of the alumina coating with the 5A zeolite pores was related with and facilely adjusted by the coating thickness. For the first time, organic molecules with sub-0.01 nm size differences were effectively distinguished via appropriate misalignment. This novel concept may have great potential to fill the pore size gaps of the zeolite family and realize size-selective adsorption separation.

444791 Dissolution and Precipitation of Minerals Under Different Alkali Concentrations in a Reactive CO_2 Batch System

Abstract:

The primary focus of this research is to study the dissolution and precipitation of minerals and salts from Berea sandstone and brine samples taken from the Great Salt Lake in a reactive CO_2 batch system. Various experiments were constructed to measure dissolution and precipitation reaction rates under different alkali concentrations. Changing the alkali concentrations of the Great Salt Lake brine samples produced changes in dissolution and precipitation of minerals from the Berea sandstone and the Great Salt Lake brine samples. Data from ICP-MS was used to quantify mineralogical changes in the samples and Micro-CT was used to measure the porosity change in the Berea sandstone samples. The results from these analytical instruments showed patterns of mineralogical dissolution and precipitation in the post-experimental fluid that correlate to the change in alkali concentrations.

*** Formatting Issue - Best Available Paper/Graphic ***

443365 Technical Reconstruction of Double Effect Distillation for Methanol Process

Abstract:

Abstract: Methanol is one of important basis raw materials of chemical industry and also an excellent environment-friendly fuel, which make it as a significant commodity in the global market. Distillation is an essential unit operation for purified production of methanol. Product quality, production cost, environmental impact and energy consumption are largely related to distillation process.

Our research effort, under this circumstance, has been devoted since 2002 to energetically optimize 4-column methanol distillation system that is widely adopted in methanol units from coal, with reduced cycling water requirement. This patented process technology saw its industrial application in Chinese largest single unit of 1800Kt/a of Baofeng Group located at North China, with the lowest steam consumption of 0.9/1 t(steam)/t(methanol) compared with normally 25% higher for both simulative case study and on-site operation. Methanol product is in accordance with O-M-232G and the recycling water is reduced significantly from 67 t(CW)/t(methanol) to 52 t(CW)/t(methanol).

The flow diagram of this advanced double effect distillation process is showed in Fig.1. Its advantages arise for limitations found in current units: (1) heat exchange network failed to fully recover energy causing large CW flow to remove extra heat; (2) for pressured column, bubble point reflux need larger reflux ratio than the sub-cooled; (3) the heat of distillate from pressured column was normally ignored; (4) the low concentration of methanol of feed inlet to recovery column leads to higher reflux radio, responsible for unsatisfied unit energy efficiency.

Fig.1 The double effect distillation process.

(Column1.Light ends column; Column 2.Pressured column;

Column 3.Atmospheric column; Column4.Recovery column)

As a solution, a new flow diagram is built up and showed in Fig.2. (1) HEN totally reorganized (not completely show); (2) sub-cooled reflux to pressured column caused 10% reduction to the reflux ratio, economizing steam consumption; (3) top product of pressured column give heat to crude methanol before feeding light ends column; (4) change product and by-product position of recovery column to give efficient unit energy consumption.

Fig.2 The reconstruction of distillation process.

(Column1.Light ends column; Column 2.Pressured column;

Column 3.Atmospheric column; Column4.Recovery column)

It is estimated that the production of methanol will be over 60 million tons in 2014. If this reconstruction could be applied in the whole methanol industry, 1.3 million tons of standard coal would have been saved. This is equivalent to reduce 5 million tons emission of carbon dioxide, so the environmental impact can be largely decreased. Field operation shows reconstruction perfect substitution worth to spread to the whole methanol industry, energetically efficient and low environmentally impacting.

444353 A Combined Rheology and Stability Study on Emulsified Acid for High Temperature Applications

Abstract:

HCl emulsion has been a popular formulation to stimulate carbonate reservoirs due to the various advantages that emulsified HCl has over ordinary HCl; including but not limited to deeper penetration and less corrosion damages. Emulsified HCl has even more potential advantages if utilized properly. The reaction kinetics, stability and rheology of emulsified HCl plays a significant role on achieving successful acidizing treatment.

The objective of this paper is to experimentally study the factors affecting the stability of emulsified HCl. The emulsion was prepared using 15wt% HCl and diesel with cationic emulsifier. The acid solution-to-diesel ratio was 70:30. The effect of stirring intensity, concentration of the emulsifier, addition of corrosion inhibitor, temperature, addition rate of acid phase to oil phase and mixing time of the emulsion have been studied. Stirring intensity was varied from 600 to 1500 rpm. Emulsifier dosage ranged between 5 to 20 gpt. Temperature used for heating, to mimic reservoir condition, was as high as 120°C.

The results showed that there is an optimum condition for emulsifier dosage and for mixing time of the emulsion. The optimum emulsifier dosage was found to be 10 gpt with 5 minutes of mixing time. As for the addition rate of acid phase to oil phase, 1.4 ml/min of drop-wise addition was found to give a very stable emulsion. The effect of stirring intensity on the emulsion stability was proportional, the higher the rpm speed the better the stability. Corrosion inhibitor, on the other hand, played a role as a destabilizer. With regards to heating temperature, it was found out that the higher the temperature, the faster the emulsion breaks.

The novelty of this study is the comprehensive approach in combining stablity and rheology of emulsified acid for high temperature applications. The particle size analysis and solution morphology help in better understanding HCl emulsion and how it works.

443503 A Computational Analysis of Fire Scene for Safety in a Compartment

Abstract:

This article details numerical prediction of fire spread carried out to study the distribution of carbon monoxide concentration, smoke dispersion, soot concentration and temperature in laboratory compartment, using a commercial fire modeling software tool. The heat release rate curve predicted from the simulation was compared with literature values carried out in a road tunnel. The time variations of carbon monoxide transportation, smoke concentration and temperature spread in compartment room are reported to evaluate the effect on humans. The soot concentration was described and used throughout this study as an indicator of safety in the compartment room. The effects of vent dimension and ventilation capacity were analyzed by smoke detector response times. This work provides improvements derived from reinterpretation of actual fire scene considering the ability of mechanical exhaust system and ventilation in compartment room to mitigate the hazard of fire.

444359 Development of Novel Emulsified Acid Using Waste Oil

Abstract:

In carbonate reservoirs, hydrochloric acid (HCl) is commonly used for stimulation purposes. The main disadvantage for using HCl is its high and rapid reactivity allowing for less control on acid reaction with formation. In addition, HCl may react with the tubing causing severe corrosion and damages. Several slow-reacting acid systems, though HCl-based, have been developed to retard the HCl reaction and control its reactivity to allow deeper penetration in carbonate reservoirs before it breaks out and reacts with formation rock. However, the selection of the hydrocarbon phase of emulsified acid is critical to the creation and stability of the emulsified acid system. Diesel oil, crude oil and solvents such as xylene have been reported to be successful in the field.

In this work, alternative and less expensive source of oil is developed to replace the current practices. Emulsified acid is a water-in-oil emulsion where oil is the continuous phase and acid is the dispersed one. Industrial and filtered waste oil was used to prepare the emulsified acid of this work. Extensive work has been carried out to study the chemical composition of the waste oil, namely SARA and to select the proper emulsifier. Emulsions were prepared using 15% HCl with emulsifier concentration ranging from 0.5 to 2.0 vol% and 70:30 of acid-to-oil ratio. Droplet size distributions were measured and used to benchmark with other emulsions. Emphasis was given to the stability of the emulsion especially at high temperatures. In order to apprehend the effect of corrosion inhibitor on the emulsion stability, its concentration was varied (up to 0.3 vol%). Later, rheological analyses of the waste oil emulsion were compared to the commonly used diesel-based emulsified acid.

Results showed the potential use of waste oil to prepare high temperature emulsified acid (up to 100°C). The addition of corrosion inhibitor to the emulsified acid caused a negative effect on stability resulting in a less stable emulsion. On the other hand, addition of cationic emulsifier helped to formulate a more stable emulsion. Apart from that, the optimum concentration of the emulsifier was at 2.0 vol%. The rheological data confirms the stability of the new emulsified formulation at higher temperatures. This paper summarizes the findings of using waste oil emulsion and recommends it for field applications. The newly proposed emulsified acid introduces waste oil as a replacement of diesel.

The novelty of using waste oil will help to significantly reduce the expenses of formulating emulsified acid. Since waste oil is basically waste it has a good impact on the environment, that being said, instead of throwing it, the industry will be able to recycle it.

443764 Optimization of Heat Exchanger Network in Steam Assisted Gravity Drainage Oil Sands Surface Facilities

Abstract:

Steam assisted gravity drainage (SAGD) is the most viable and environmentally safe recovery technology for extracting heavy oil and bitumen. It is a thermal in-situ production process where parallel wells are drilled horizontally into an underground bitumen reservoir. Steam is produced at the central processing facility (CPF) and injected into the reservoir through the wells. The steam heats the bitumen to a point where gravity allows it to flow down to the lower well where the mixture of bitumen and water is then pumped to the surface. The water and bitumen are separated at the CPF. Typically, more than 90% of the energy consumed in producing a barrel of bitumen using SAGD is used to generate the steam that is injected into the bitumen reservoir. Fossil fuels is employed in the process, which causes GHG (greenhouse gas) emissions. The usage of hot and cold utility is reduced by increased heat exchange and optimization of the heat exchanger network, which results in reduction of GHG emissions. In this work, pinch analysis is performed to minimize energy consumption.

443855 Modeling the Thermal Behaviors of an Ultracapacitor Module

Abstract:

Ultracapacitors, also known as supercapacitors, have the potential to meet the increasing power requirements of the sustainable energy systems for the grid-scale energy storage systems and electrified vehicles. As compared to batteries, ultracapacitors offer a higher power density, higher efficiency, and longer shelf and cycle life. Because the performance of an ultracapacitor depends on temperature, it is important to calculate accurately the thermal behavior of a single ultracapacitor cell and the module composed of multiple ultracapacitor cells for the efficient and reliable systems integration of an ultracapacitor in grid and transport applications.

In this work, modeling is performed to study the effect of the layout configugation of the ultracapacitor cells on the thermal behaviors of an ultracapacitor module. The ultracapacitor module is subject to the charge/discharge cycling with constant-current between the nominal voltage of the ultracapacitor module and the half of this voltage. The charge/discharge current values examined are 50, 100, 150, and 200 A. The validation of the modeling approach is provided through the comparison of the modeling results with the experimental measurements.

443856 Modeling the Electrolyte Flow Behaviors of a Zinc/Bromine Flow Battery Module

Abstract:

The zinc/bromine (Zn/Br_2) flow battery is a promising rechargeable power source for energy storage systems because of its inherent chemical simplicity, high degree of reversibility at the electrodes, good energy density, and abundant low-cost materials. The Zn/Br_2 flow battery system, intended for the grid-scale energy storage system, is to be made in modular form. A Zn/Br_2 flow battery module is composed of the multiple stacks consisting of bipolar carbon electrodes. Since the stacks require electrolyte circulation on both the negative and positive sides, two plumbing loops are provided. The flow loops contain an electrolyte storage tank and a circulation pump. A heat exchanger may be included for temperature control in the flow loop. The design of the electrolyte circulation plumbing determines the distributions of electrolyte flow and pressure within the stacks and the efficiency of the stack is dependent on the flow distribution throughout the stack. The modeling of the distributions of electrolyte flow and pressure within the module is needed to optimize the plumbing for electrolyte circulation and thereby to improve the performance of the module. In this work, modeling is performed to study the effect of the layout configuration of the plumbing for electrolyte circulation on the distributions of electrolyte flow and pressure within a 50 kWh Zn/Br_2 flow battery module composed of four stacks. It is seen that the plumbing layout for electrolyte circulation can have a significant effect on the flow rate distribution of electrolyte through the stacks.

443974 Observer-Based Model Predictive Control for Time-Delay Systems

Abstract:

Model predictive control is one of the most widely used advanced process control technologies in industry.[1] It has many capabilities such as dealing with input/state constrains and uncertainties. Observer-based model predictive control was developed to estimate states information when the states are not measurable directly. Many chemical processes have time delays in either states or measurements, such as polymer production and some bioprocesses. [2] In this presentation, we are proposing an unknown input observer based model predictive control algorithm for time-delay systems. Unknown input observer is an observer design to eliminate certain types of unknown inputs in the system. Stability of unknown input observer-based model predictive control without time delay was provided by our previous study. Terminal cost and terminal region are designed for a class of non-linear time delay systems. The close-loop stability of the controller is provided. [3][4]

An example of bioreactor is provided which has uncertain time delays in measurements but with a known upper bound. Two measurements are provided which are online biomass and offline ethanol measurements. Model-plant mismatch is considered as unknown inputs in the system. State estimation and control are calculated using the observer and an optimization algorithm.

Reference:

[1] Marcus Reble et al., Model predictive control of constrained non-linear time-delay systems. IMA Journal of Mathematical Control and Information, 2010, 1-19

[2] Seung Cheol Jeong and PooGyeon Park, Constrained MPC algorithm for uncertainty time-varying systems with state delay, IEEE TRANSACTIONS ON AUTOMATIC CONTROL, 50, 2005, 257-262

[3] Mayuresh Kothare et al., Robust constrained model predictive control using linear matrix inequalities. Automatica, 32, 1996, 1361-1379

[4] Wook Hyun Kwon et al., A simple receding horizon control for state delayed systems and its stability criterion, Journal of Process Control, 13, 2003, 539-551

444044 Organic Solar Cells

Abstract:

This paper describes highlights the need of next-generation "organic solar cells". There is a continuing need for organic solar cells and arrays with very high specific power densities (1000-5000 kW/kg) for generating power in a new generation of ultra light space payloads example in NASA. An emerging technology with promise to meet these ambitious goals are organic solar cells based on very thin films of discrete layers. Theoretical efficiencies for multi junction solar cells are more ~70% as compared to ~30% for single junction devices. There is continuous development going on high mobility, band gap engineered organic donors and acceptors, as well as development of three junction devices and scaled up devices on lightweight flexible polymer substrates. The main interest in this paper is organic semi-conducting crystal: Pentacene, which has photo-conducting states. Organic solar cells manufactured today have low efficiency other than their stability issues. The efficiency of organic solar cells can be improved by either designing polymer materials that absorbs more of solar spectrum or by minimizing the charge collection efficiency by controlling the morphology of polymer films. Once fully developed, organic solar cells would revolutionary power the industry due to their flexibility and minimal weight

444055 A Refined Non-Random Two-Liquid Segment Activity Coefficient Model for Solubility Modeling

Abstract:

Accurate correlation and robust prediction for complex molecule solubility or solid-liquid equilibrium (SLE) in solvents and solvent mixtures have drawn much interest recently [Tanveer et al., *Chem. Eng. Prog.* **2014**, 110 (9), 37-47]. Solubility models such as the Non-Random Two Liquid Segment Activity Coefficient (NRTL-SAC) model [Chen and Song, *Ind. Eng. Chem. Res.* **2004**, 43, 8354-8362] are now routinely used in the industries. However, an important unfilled need is to model the solubility in oligomers/polymers/surfactants such as asphaltene nanoaggregates in petroleum fluids and drug molecules in lipid-based or solid dispersion formulation [Tung, *Org. Proc. Res. Dev.* **2013**, 17, 445-454]. This paper presents a refinement to the very successful NRTL-SAC model to better account for the contribution due to size differences of molecules. Specifically, the Flory-Huggins combinatorial entropy expression for conceptual segments in the model is substituted with the Staverman-Guggenheim expression for real molecules. We present the rationale for the refined model, the updated conceptual segment numbers for common solvents, and the newly identified conceptual segment numbers for common drug excipients. The model performance is illustrated by prediction of drug molecule solubility in various lipid solvents and solvent mixtures, compared to experimental data and previous work by other models. The refined NRTL-SAC model shows enhanced capability to correlate and predict drug molecule solubility in common solvents and excipients used in drug formulation.

443640 Preparation and Characterization of (*FURCRAEA spp*) as Green Support of Iron Nanostructured Catalyst

Abstract:

PREPARATION AND CHARACTERIZATION OF (*FURCRAEA spp*) AS "GREEN" SUPPORT OF IRON NANOSTRUCTURED CATALYST

Karen G. Bastidas G[1], Anamaria Barrera B[1], Cesar A. Sierra A[2], Hugo R. Zea R[3]

[1] Department of Chemical and Environmental Engineering. National University of Colombia (Bogota).

[2] Associate professor. Director of the Research Group of Macromolecules. National University of Colombia (Bogota).

[3] Associate professor. Director of the Research Group of Materials, Catalysis and Environment. National University of Colombia (Bogotá).

Abstract.

The availability of natural water resources has declined over time, becoming one of the biggest problems today. Environmental risk in Colombia's rivers has increase due to the widespread use of harmful chemicals in a variety of processes, ranging from illegal mineral extraction to industry wastewater, e.g. Medellin's river, due to its strategic location along a textile industrial corridor might be subject to dye contamination; the gold and platinum mining zones in Choco and Antioquia regions are being heavily affected by the use of mercury in the extraction processes. In 2009, 520 tons of mercury were imported by miners working in this area, amount that exceeds several times the mercury used by regulated industry in previous years (Ministry of Mines and Energy of Colombia, 2007). Not only the environmental risk is a concern but also the economic conditions of many low-income families whose livelihood in many cases depends exclusively of the of the mining.

Due to the issues raised above, the main goal of this study is to research, develop and implement innovative technologies in the synthesis of highly efficient, low cost and environmentally friendly iron based catalysts capable of degradate pollutants, ensuring operational stability. A native fiber (Fique, FURCRAEA spp) has been selected as a solid support for these catalysts due to its abundance, low cost and special physical and chemical characteristics.

The combination of cellulosic fibers with iron oxide nanoparticles could provide exceptional biodegradable composite materials for the treatment of organic dyes and heavy metals present in wastewater. Likewise, the use of iron particles immobilized on fique fibers allows an easy separation catalyst-treated water, turning the biocatalyst in highly recyclable and reusable material.

Fique fiber was initially characterized by determining the percentage of ashes, total lignin, soluble and insoluble lignin, cellulose, hemicellulose, total solids and humidity, obtaining 0.94, 22.58, 0.5, 22.8, 36.3, 27.5 0.701 and 2.1%, respectively. Additionally, the isoelectric point (pzc), acidic and basic sites were measured.

Then, the fiber was subjected to a pretreatment in order to expose more active sites which on time serve as nucleation points for the deposition of nanoparticles. This was done by means of an ultrasound-assisted procedure, varying the sonication time (30, 60 and 90 minutes). The effect of the sonication was followed by measuring the contents of lignin, cellulose and hemicellulose within the fiber after each time. 60 minutes of sonication was found to be most beneficial time since the amount of lignin and hemicellulose present in the fiber was reduced, while conversely the amount of crystalline cellulose exposed increases.

In order to improve the chemical ability of the nucleation sites to interact with iron and then to form stable and size controllable nanoparticles, a cationization procedure was performed on the fiber; this was achieved by the exposure of the fiber to acidic and basic conditions promoting cationization. Three cationization times (1, 2 and 3 hours) were studied and their effect on the fiber chemical surface was evaluated using FTIR-ATR and XRD analysis. The results showed 3 hours as the optimal cationizing time.

Subsequent to the cationization of the fiber surface, a procedure of ionic/cationic exchange was implemented by exposing the fiber to a solution of iron salts in order to replace the cations in the fiber surface and anchoring the iron nanoparticles. Different conditions were studied in order to synthetize zero valent iron nanoparticles (nZVI) and others with different oxidation states. Finally, to track the impregnation process on the fiber, samples from the precursor solution were taken at different times and its iron concentration was determined by atomic absorption spectrophotometry (AAS). Complementary characterization performed by X-ray diffraction, BET surface area and electronic microscopy, among others.

Key words: Natural support (fique), iron nanoparticles, bionanocatalyst.

444057 Explaining the Paradox of Asphaltene Precipitation with n-Alkanes

Abstract:

Explaining the paradox of asphaltene precipitation with n-alkanes

Md Rashedul Islam and Chau-Chyun Chen
Department of Chemical Engineering, Texas Tech University

Abstract

Paraffins, specifically alkanes, promotes asphaltene precipitation when they are blended with asphaltene-containing petroleum fluids. The alkane/oil volume ratio required for onset of asphaltene precipitation increases with the carbon number and yields a maxima. Experimental study indicates that the maxima lays at C7[1] or between C8 -- C10 [2]. However, precipitation yield decreases asymptotically with increase of the carbon number even if excessively high alkane/oil volume ratio is used [3]. For example, n-pentane be a better solvent for asphaltene than n-hexane in regard of both onset of precipitation and precipitation yield. However, high carbon number alkane beyond the maxima, n-hexadecane for instance, can be a better solvent at onset of precipitation point and a poorer solvent in regard of precipitation yield than n-pentadecane. A thermodynamic explanation for this paradox is necessarily important to find the compatibility of bitumen or crude oil with paraffinic hydrocarbons. In this work, this paradox is explained by the aggregation thermodynamics of asphaltene precipitation [4]. Both the enthalpy and entropy of mixing are investigated to explain the effect of paraffin carbon number on asphaltene precipitation. Athabasca and cold lake bitumen samples are studied.

References

[1] G. Hotier and M. Robin, "Action de divers diluants sur les produits pÃ©troliers lourds: mesure, interprÃ©tation et prÃ©vision de la floculation des asphalténes," Oil & Gas Science and Technology, vol. 38, pp. 101-120, 1983.

[2] I. A. Wiehe, H. W. Yarranton, K. Akbarzadeh, P. M. Rahimi, and A. Teclemariam, "The paradox of asphaltene precipitation with normal paraffins," Energy & fuels, vol. 19, pp. 1261-1267, 2005.

[3] D. L. Mitchell and J. G. Speight, "The solubility of asphaltenes in hydrocarbon solvents," Fuel, vol. 52, pp. 149-152, 1973.

[4] M. Wang, Y. Hao, M. R. Islam, and C. C. Chen, "Aggregation Thermodynamics for Asphaltene Precipitation," AIChE Journal, Submitted.

Using a Large Data Bank to Develop a Robust and Simple Model to Estimate the Saturation Pressure of Oil Reservoirs

By

Hassan Alkandari[1], Adel Elsharkawy[1], Adel Malallah, Osamah Alomair

Petroleum Engineering Department, College of Engineering & Petroleum
Kuwait University
Po Box 5969 Safat 13060-Kuwait

Abstract

Measurements of saturation pressure are crucial for all hydrocarbon reservoir fluids. Below the crude oil saturation pressure gas reaches a critical saturation; then, a two phase flow occurs and results in decreasing oil production and recovery. To maximize oil production and recovery, the reservoir pressure has to be maintained closer to the original saturation pressure. This pressure is normally measured using bottom-hole samples or surface recombined samples of oil and gas. Occasionally, the samples become unavailable and the pressure needs to be estimated using computational methods.

In this study, a large data bank of 231 crude oil compositions and saturation pressure measurements including literature data, unpublished data, and newly measured data were used to develop two empirical models to predict the saturation pressure of a variety of crude oils if the oil sample is unavailable or the experimental measurements are unreliable. The first proposed model utilizes the extended compositions of hydrocarbons up to the heptane plus fraction in addition to non-hydrocarbons. The second model uses the lumped compositions of light, intermediate, and heavy components in addition to non-hydrocarbon components as the input variables. The lumped model has the advantage of using fewer input parameters while maintaining the thermodynamics. The accuracy and validity of both models to calculate the saturation pressure for volatile oils, black oils, and heavy oils are presented using several compositional data. The models performance is also compared to the Peng-Robinson equation-of-state (PR-EOS), and the Soave-Redlich-Kwong equation-of-state (SRK-EOS) as well as all published methods that use compositions as input variables. The comparison indicates that the proposed models are much simpler and more accurate than the other computational methods. The proposed models treat the heptane-plus fraction as a single component, thus eliminating the splitting and characterizing the plus fractions and the binary interaction parameters needed for the EOS's calculations.

Key words: Saturation pressure (Ps), Bubble-point pressure (BPP), Equation of state (EOS), Correlations, and PVT Data

[1] Corresponding Authors: Hassan Alkandari , Hassan@hkandari.com, Adel Elsharkawy, adel.elsharkawy@ku.edu.kw

Introduction:

Reservoir fluid properties are an essential source of information for all calculations on hydrocarbon reserves. Accurate estimation of the pressure-volume-temperature (PVT) measurement is crucial. Of all the PVT properties, saturation pressure (Ps) is considered the most important parameter. It is well known that gas starts to flow once the reservoir pressure drops below the bubble point pressure and the gas saturation reaches a critical value. To minimize the gas flow and maximize the oil recovery, it is very important to maintain the reservoir pressure close to the original bubble point pressure.

Although the PVT properties are experimentally measured in laboratories and provide reliable results, this process is expensive, time consuming, and sometimes unavailable; therefore, measured field data (separator gas oil ratio, separator pressure, stock-tank oil specific gravity, and reservoir temperature) or reservoir fluid compositions are used to estimate the bubble point pressure.

There are two methods in the petroleum industry to calculate the saturation pressure of crude oil when experimental is unavailable; either using production data or compositional data.

Methods Based on Production Data (Composition is unavailable)

Numerous attempts have been made since the 1940s to develop correlations to estimate the saturation pressure from production data, such as producing gas oil ratio, oil gravity, gas gravity, reservoir pressure and temperature. Table 1 presents summary of all the methods that use production data to estimate the saturation pressure.

Methods based on Compositional data

Many authors have reported that the application of any correlation in other regions results in serious errors. These conclusions arise from the fact that regional crude oils represent a physical-chemical trend; therefore, all attempts to develop a general correlation using field measured parameters to estimate the saturation pressure from a large data bank have failed. For this reason, Elsharkawy (2003) made the first attempt to correlate the saturation pressure to the reservoir fluid composition instead of using the production data. In recent years, there have been several attempts to estimate the saturation pressure of oil reservoirs. These methods are reported in Table 1.

Elsharkawy (2003) presented a correlation of the bubble point pressure for the oil reservoir using the same input data needed for the equation-of-state (EOS) calculations, i.e., the compositional data and the reservoir temperature. Elsharkawy's model is based on the saturation pressure measurements of 60 crude oil samples from the Middle East and 75 collected from the literature. He demonstrates that the calculation of the saturation pressure depends on the type of EOS, the number of pseudo components, the method used to characterize the pseudo-components and the binary interaction number. Elsharkawy compared the performance of his model to the SRK-EOS and PR-EOS for black oils, volatile oils, heavy oils, and synthetic oils. He indicated that his model eliminates splitting the heavy plus fraction, characterizing the pseudo-components and inclusion of the binary interaction parameters needed for the EOS calculations.

Later, Tale et al. (2009) made the second attempt to model the saturation pressure of crude oil using 132 data samples published by Elsharkawy (2003). Tale's model, which is based on the alternating conditional expectation (ACE), has a set of 14 equations with 42 constants. The average absolute relative error (AARE) reported by Tale et al. indicates that the model did not perform well with the selected data. The ACE model has an AARE of 12.3% compared with AARE of 6.7% by Elsharkawy (2003) and 8.1% AARE for the SRK-EOS.

AlQuraishi (2009) attempted to model the saturation pressure using Elsharkawy's (2003) data. The model was developed using linear genetic programming. He correlated the saturation pressure to the formation temperature, methane content, and heptanes plus molecular weight. Thus, AlQuraishi's model neglected the effect of the intermediate and heptane plus content on the saturation pressure. The error analysis reported by AlQuraishi indicates that his model has a 5.8% error compared to 7.7% by Elsharkawy, 9.9% using the SRK-EOS, and 10.1% using the PR-EOS.

Bandyopadhyay and Sharma (2011) presented a model to predict the bubble point pressure, using temperature and fluid composition. They used 129 experimental and literature data originally published by Elsharkawy (2003) to develop their model. The model introduced a temperature interaction parameter that depends on compositions. The proposed model has a set of 24 coefficients and the estimated bubble point pressure is calculated as the sum of the series calculations. They reported an average absolute error (AAR) of 7.83% using their model compared to 8.3% using Elsharkawy's simple model (2003), 10.15% using SRK-EOS, and 10.58% using the PR-EOS.

Khamehchi et al. (2011) used an intelligent proxy simulator (IPS), which is based on an adaptive neural fuzzy inference system (ANFIS) and a structured artificial neural network (ANN) to predict the bubble point pressure and asphaltene onset pressure. The model uses a reservoir fluid composition and temperature to predict the Ps. It is not clear how the Ps is calculated from the proposed method, and they did not present the accuracy of the proposed method compared to the other published methods.

Farasat et al. (2013) presented a new mathematical approach to calculate the crude oil saturation pressure as a function of the temperature and the reservoir fluid composition. The model developed by Farasat et al used 130 experimental data, which were originally published by Elsharkawy (2003). Farasat used a support vector machine (SVM) to model the saturation pressure. They did not clearly demonstrate how to calculate the saturation pressure from the SVM model. Moreover, they did not indicate the overall average absolute deviation (AAD) for their model compared to other published models and both the PR-EOS and SRK-EOS for the 130 data.

Ahmadi et al. (2014) also presented an ANN technique called gene expression programming (GEP) to determine the bubble point pressure of the oil samples. Similar to Farasat et al. (2013), Ahmadi's model used most of the data previously published by Elsharkawy (2003) to construct a non-linear equation using reservoir fluid compositions and temperatures as input parameters. The proposed model is highly complicated and consists of 8 equations that have several constants compared to the one simple equation proposed by Elsharkawy (2003). They compared the performance of their proposed model to the SRK-EOS and PR-EOS; however, their paper does not mention how the heptanes plus fraction were treated and characterized for the EOS calculations. They reported an AAD of 4% using the proposed model compared to 8% using Elsharkawy's model (2003), 10.9% by PR-EOS, 10.5% by SRK-EOS, and 15% by Bandyopadhyay and Sharma (2011).

Gholami et al. (2014) estimated the bubble point pressure using a support vector regression SVR and a supervised learning algorithm SLT based on the statistical learning theory, and then the ACE was used. The model used reservoir fluid compositions and temperature to estimate the bubble point pressure. Gholami et al. used the same data originally published by Elsharkawy (2003). They presented three attempts based on genetic programming to correlate the bubble point pressure: SVR, ACE, and power-law committee machine (PLCM). The ACE-based method resulted in a set of three equations with a matrix of [8 by14] constants. Gholami et al. compared the accuracy of their model to the SRK-EOS and PR-EOS predictions reported by Elsharkawy (2003). Based on their AARD analysis, the PLCM model has the lowest errors followed by the SVR model, Elsharkawy's model, ACE model, PR-EOS, and SRK-EOS.

Lately, Jarrahian et al (2015) presented two empirical models to estimate the saturation pressure of black oils. The first model is based on production information. The second model is based on compositional information mostly from Elsharkawy (2003) and Heidaryan and Moghadasi (2011). They have not presented comparison of the accuracy of their model to Farasat (2013) model, Ahmadi (2014) model, and Gholami (2014) model.

This paper is concerned with the estimation of saturation pressure using compositional data. Therefore, the objective is to develop a robust yet simple model to estimate the saturation pressure for various types of reservoir oils using a larger data bank than that used by other published studies. The accuracy of the proposed models is compared to previously published models as well as SRK-EOS and PR-EOS.

Data bank

This study uses a large data set of 231 crude oil compositions and saturation pressure measurements. The data set consists of 130 data sets originally reported by Elsharkawy (2003); in addition to others collected from literature (Jacopy, 1958; Vogel, 1980; Coats, 1986; Droham, 1980; Dansh, 1992; Moharam, 1996; Rosenegger, 1999); 94 new data sets that have not been utilized before; as well as 7 newly measured samples at the Petroleum Fluid Research Center (PFRC) in Kuwait University. These data represents a wide range of compositions, such as volatile oils, black oils, and heavy oils. The data cover a pressure range of 537 to 5065 psia and a temperature range of 58 to 319 °F, crude oil compositions from C_1 to C_{7+}, non-hydrocarbons, such as N_2, H_2S, and CO_2, and the molecular weight and the specific gravity of the heptane plus fraction.

Data screening and classification

The compositions and the saturation pressure measurements in the data bank are subjected to a meticulous screening process to check the validity of the data. Our screening criteria uses the following steps: the sum of the molar compositions of all the components must add up to one, otherwise, the sample is eliminated. Samples with a missing molecular weight or specific gravity of the plus fraction are also eliminated.

After filtering, the data were reduced to 214 and divided into 179 measurements used to develop the new models, and 35 data sets are used for the blind test of the robustness of the proposed models. Statistical analysis of the input variables of the data set used in training and testing the extended and the lumped models are illustrated in Tables 2.

Several of the reported compositional data have extended analysis up to (C_{20+}). The following techniques are used for lumping or regrouping the extended analysis into single pseudo-fractions (C_{7+}). The mole fraction (X_i) of the individual components of the extended analysis is equal to the plus fraction (Ahmad, 2011).

$$X_{C_{7+}} = \sum_{i=7}^{n} X_i \qquad i=7, 8 \ldots n \qquad (1)$$

The average molecular weight (Mw) of the individual component of the extended analysis is equal to the molecular weight of the plus fraction.

$$Mw_{C_{7+}} = \sum_{i=7}^{n} \frac{X_i Mw_i}{X_{C_{7+}}} \qquad i=7, 8 \ldots n \qquad (2)$$

The density of the total plus-fraction can be calculated as follows:

$$\gamma_{C_{7+}} = \frac{X_{C_{7+}} Mw_{C_{7+}}}{\sum_{i=7}^{n} \frac{X_i Mw_i}{\gamma_i}} \qquad i=7, 8 \ldots n \qquad (3)$$

It is important to note the significance of the density of plus fraction because its density determines its paraffinic naphthenic aromatic (PNA) content.

Equation-of-state calculations

The EOS calculation of the saturation pressure is compared with the newly developed models as well as the other methods. The two EOS calculations considered in this paper are the ones most considered and widely used in the petroleum industry to evaluate the volumetric and phase behavior of reservoir fluid properties: the Soave-Redlich-Kwong EOS (SRK-EOS) and the Peng-Robinson EOS (PR-EOS). It should be noted that the accuracy of any EOS calculation of the saturation pressure largely depends on the number of pseudo-components and the splitting schemes used to divide the heptane plus fraction; as well as the characterization and calculation of the critical properties of the pseudo-components and the binary interaction parameters. Therefore, the two EOS calculations of the saturation pressure are not the same, even though the number of pseudo-components, characterization of the pseudo-components, and binary interaction parameters were the same.

Model development

As previously stated, the objective of this work is to provide the petroleum reservoir engineer with a simple yet accurate tool to estimate the saturation pressure for a variety of crude oils using compositional data when experimental is unavailable. Two models are presented in this study to calculate the saturation pressure of the crude oil samples. The first model uses all input parameters, such as the molar compositions of the hydrocarbon and non-hydrocarbon components similar to the EOS models. This model is referred to as the extended model. The second model uses fewer input parameters in which some of the hydrocarbon components have been lumped into a single group; this is designated as the lumped model. Both models are simpler than all published correlations, physically correct, and have an acceptable accuracy compared with the well-known EOS.

Figure 1 shows a correlation matrix of all the data available under consideration. This figure also indicates the relative importance of each of the input variables into the first proposed model. It agrees with most published correlations and the EOS that the methane content, heptane plus content and temperatures have the greatest impact on the saturation pressure followed by the intermediate content. The data understudy indicates that the non-hydrocarbon has the smallest impact. This result is simply a reflection of the molar contribution of each component comprising the crude oil composition.

Model I - Extended Model

The extended model uses 13 variables comprising the molar compositions of the pure components, heptane plus properties, and reservoir temperatures, which are similar to the input parameters for the EOS. Figure 2 shows the correlation coefficient graph of all the input parameters in the extended model. Thus, the elimination of the intermediate effect has a negative impact on the accuracy of the predictive model. Our proposed extended model has the following form:

$$Ps = 124.72N_2 + 17.57CO_2 + 22.55H_2S + 64.22C_1 - 9.80C_2 - 52.49C_3 + 6.16C_4 - 19.22C_5 - 23.63C_6 - 21.43C_{7+} + 435.31\gamma_{C7+} + 1.14Mw_{C7+} + 4.29T \quad (4)$$

Where Ps is the saturation pressure in psia, T is the temperature in °F, and the hydrocarbon and non-hydrocarbon compositions are expressed in mole.

Model II - Lumped Model

The second model is much simpler and uses only 7 -variables (the lumped model). Several lumping schemes are adapted in the petroleum industry for phase behavior modeling and reservoir simulation of the petroleum reservoir fluids (Ahmed, 2011). Hydrocarbon components are grouped into a pseudo fraction based on their molecular weight (Whitson, 1980), the similarity of their physical-chemical properties (Lee, 1979), their carbon number, their true boiling point or their number of families based on the chemical structure (Becker et al., 2015). (Emera and Samara, 2005) proposed lumping the hydrocarbon components into volatile (C_1 & N_2), intermediate (CO_2, H_2S, C_2 to C_4) and heavy categories (C_{5+}) to calculate the minimum miscibility pressure. (Duan et al., 2013) proposed lumping several components into a single carbon number (SCN) with predetermined values for properties, such as critical pressure, temperature, and eccentric factor. Others have proposed optimal lumping schemes based on the objectives (Lin et al., 2008). In lumped model, $C_{1+}N_2$, C_2, CO_2, and H_2S, the intermediate (C_3 to C_6), heptanes plus (C_{7+}) and its properties, and temperature are the input variable. The correlation matrixes of the lumped components are provided in Figure 3. Even though the non-hydrocarbon has the least effect their contribution is important, especially the H_2S in the case of sour crudes. Both models presented by AlQuraishi (2009) and Bandyopadhyay and Sharma (2011) have eliminated the effect of H_2S.

The proposed model has the following form:

$$Ps= 82.115*(C_1+N2)-11.635*C_2+39.158*CO_2+ 38.244*H_2S -1.217*(C_3—C_6) - 890.701*(C_{7+}/MW_{7+}*\gamma_{7+}) + 4.217 *T-1042 \quad (5)$$

Where Ps is the saturation pressure in psia; and T is the temperature in °F.

Result and discussion

Model accuracy

The accuracy of the two proposed models used to calculate the saturation pressures of the previously described crude oil samples is shown in Table 3. This table shows reports the average relative error, average absolute relative error standard deviation and the correlation coefficients for all the methods considered in this study. Table 3 also reports the accuracy of the PR-EOS, SRK-EOS, Elsharkawy's model, AlQuraishi's model, and Bandoypandyay and Sharma's model and Jarrahian model. The methods proposed by Gholami et al., Ahmadi et al., and Jarrahian et al are included in Table 3, even though they produce a unreasonable estimate of the saturation pressures, as explained in the appendix.

This table indicates that the extended and the lumped models have an (AARE) in the order of 8% and 8.5%, respectively. It should be noted that the complications in the Bandoypandyay and Sharma model did not improve the accuracy of the saturation pressure, because it had a large error. It is also noted that Jarrahian et al model produces an average error of 35%. Compared with the EOS, the lumped model has a few constants and eliminates the splitting, characterizing the pseudo-components and binary interaction parameters necessary for the EOS calculations. The lumped model as well as the extended model are very simple and require one-step calculations compared with all the other methods reported in Table 3.

A cross-plot of the calculated saturation pressure using the extended model based on the training data versus the measured data is illustrated in Figure 4A. This figure indicates that most of the data falls on the unit slope line, and there is no over-fit. This trend is supported by the high correlation of regression of the model (97.6%) presented in Table 3. Figure 4B shows a cross-plot of the testing data (28 from literature and 7 new measured in PFRC at Kuwait University). The extended model indicates an excellent agreement with all of the 28 data sets chosen at random, with the exception of a few data points at higher pressures. Additionally, the model has a perfect match for the 7 newly measured samples.

Figure 5A shows the performance of the lumped training-model for capturing the measured saturation pressure for most of the 179 samples. A cross-plot of the calculated saturation pressure using the lumped model based on the training data versus measured data is depicted in Figure 5B. This figure indicates that most of the data scatters falls to the unit slope line. The lumped model also has a high coefficient of regression of 97.4%, which is indicated in Table 3.

Models Validity

The accuracy and validity of the models for estimating the saturation pressure of various types of crude oil are reported in Table 4. This table illustrates the capability of the newly proposed models to predict the saturation pressure for volatile oils, black oils, and heavy oils. Table 4; provides the compositions, properties of the heptane plus fraction, temperature, and measured saturation pressures of the samples. Additionally, the calculated saturation pressure using the PR-EOS, SRK-EOS, Elsharkawy's model (2003), AlQuraishi's model (2009), and Bandoypandhyay and Sharma's model (2011), Jarrahian et al (2015); as well as the extended and lumped models are provided in Table 4. It is clear from this table that the proposed models have a wide range of applications regardless of crude oil types. We have also included information regarding the 7 newly measured data; at the PFRC, compositional, and saturation pressure. At the bottom of Table 4, we have provided the unpublished data of the compositions; and the saturation pressures at low temperatures. It is well known that the saturation pressure of a crude sample is measured at low temperatures in addition to the reservoir temperature as a method to check the quality of the sample.

Figure 6A reports the error of all the methods considered in this paper to predict the saturation pressure for volatile oils (heptane plus content below 30%). Both the extended and the lumped models have an AARE of nearly 4% followed by Bandoypandhyay and Sharma's model (2011); and Elsharkawy's model (2003); the SRK-EOS has nearly twice the AARE of the newly proposed model. AlQuraishi's model (2009) has an AARE of 12% because the model is based on methane and neglects the effect of heavy and intermediate components, whose contribution is highly significant in the reservoir and production performance of volatile crudes. Jarrhaine et al (2015) model is not suitable for volatile oils. It shows a 35% error in estimated saturation pressure.

The performance of the extended and lumped models compared to the various methods for the black oil samples ($35< C_{7+} <55\%$) is provided in Figure 6B. The newly proposed models have an AARE close to 10%; followed by Bandoypandyay and Sharma's model 14%, the SRK-EOS and PR-EOS 16%, and Elsharkawy 17%. AlQuraishi's model has an AARE of 24%. Jarrhain et al (2015) model shows a 33% error in estimated saturation pressure even though the authors claim that the model is suitable for black oils.

Figure 6C shows the error level of all the methods as well as the proposed models for the newly measured crude oil samples at Kuwait University using the PFRC. Again, Jarrhain et al (2015) model show a 40% error in estimated saturation pressure.

The performance of all published models at low temperature is illustrated in Figure 6D. The low-temperature calculation indicates that Bandoypandhyay and Sharma's model (2011) has accuracy problems and is not valid at low temperatures because the authors proposed a temperature-composition related coefficient. Also, Jarrhain et al (2015) model is not suitable at low temperature. It shows a 40% error in estimated saturation pressure.

The ability of the proposed models to capture the thermodynamic is thoroughly checked against its independent variables. The validity of the model is compared to the experimentally measured data as well as the predictions by other methods used to capture the physics of increasing the saturation pressure as a function of temperature is illustrated in Figure 7. Two main points are observed here; first, the model matches the measured saturation pressures much better than the other methods. Second, it follows the physical trend predicted by the EOS's. Figure 8 shows the model performance for capturing the effect of changing the saturation pressure as a function of the molecular weight of the heptane plus fraction compared with the SRK-EOS and PR-EOS. The extended model follows the trend of the PR-EOS in increasing the saturation pressure. Elsharkawy's model (2003) was unable to predict the correct trend as predicted by the EOS's. The model sensitivity in capturing the change in the saturation pressure as a function of increasing the methane content is shown in Figure 9. The proposed model as well predicts the correct trend of increasing the saturation pressure when increasing the methane content.

Conclusion

In this paper, two models are presented to estimate the saturation pressure of a variety of crude oils when measures are unreliable or samples are unavailable. The two models are developed using compositional data and saturation pressure measurements from a large data bank comprising literature data, unpublished data, and newly measured and reported data of 231 crude oil samples. The input parameters in the first model are the extended hydrocarbon and non-hydrocarbon molar percentages and reservoir temperatures. The second model uses lumped hydrocarbon fractions of intermediate, and heptane plus fractions in addition to its properties; and the reservoir temperature as input parameters. The accuracy of the two models has been tested and compared to that of the SRK-EOS and the PR-EOS; as well as published correlations that use compositional data as the input parameters. The comparison indicates that the two models are much more accurate than all the published methods. The validity of both models has also been tested using various types of crude oils samples and compared to experimentally measured saturation pressures and estimates using the EOS as well as the published correlations. The proposed models are physically correct and much simpler than all the published methods. The models do not require tedious computations and eliminate splitting of the plus fraction, characterization of the pseudo-components; as well as the binary interaction parameters that are required for the EOS calculations.

Acknowledgement

The authors thank the General Facility Lab at Kuwait University for their cooperation and conducting the compositional and saturation pressure measurements of the 7 new samples described in this paper using Petroleum Fluid Research Center (PFRC) project no. GE 01107.

References

Agarwal, R.K., Li, Y.K., Nghiem, L., 1990. A regression technique with dynamic parameter selection for phase – behavior matching. SPE Res. Eng. February, 115 – 120.

Ahmadi, M. A., Zendehboudi, S, James, L., Elkamel, A., Dusseault, M., Chatsiz, I., Lohi, A., 2014, New Tools to determine Bubble Point Pressure of Crude Oils: Experimental and Modeling Study, *Journal of Petroleum Science and Engineering, v.123, 204-216.*

Ahmed, T., 2011, Splitting and lumping schemes of the plus fraction, working guide to vapour-liquid phase equilibria calculations, Chapter 9, Hydrocarbon Phase Behaviour, Gulf Publishing Co. 111-126.

Al-Gathe, A., Abd-el Fattah, Kh., El-Mtwally, K., 2014, New artificial intelligent approach for bubble point pressure, Int. J innovation and Sci. Research, Vol.4, No.2, July, 121-135.

Al-Marhoun, M.A. 1985. Pressure-Volume-Temperature Correlations for Saudi Crude Oils. Presented at the SPE Middle East Oil Technical Conference and Exhibition, Bahrain, 11–14 March. SPE-13718.

Almehaideb, R.A. 1992. Improved PVT Correlations for UAE Crude Oils. Presented at the Middle East Oil Show and Conference, Bahrain, 15-18 March. SPE-37691.

Al-Najjar, H.S., Al-Soof, N.B.A., and Al-Khalisy, K.M. 1988. Correlations for Bubble-Point Pressures, Gas Oil Ratios and Formation Volume Factors for Iraqi Crude Oils. *Journal of Petroleum Research* (June 1988): 13.

AlQuraishi, A. A., 2009, Determination of crude oil saturation pressure using linear genetic programing, Energy & Fuel, 23, 884-887.

Al-Shammasi, A.A. 2001. A Review of Bubblepoint Pressure and Oil Formation Volume Factor Correlations. *SPE Res. Eval. & Eng.* 4 (2): 146-160. SPE71302.

Asgarpour, S., McLauchlin, L., Wong, D. et al. 1989. Pressure-Volume-Temperature Correlations for Western Canadian Gases and Oils. *J Can Pet Tech.* 28 (4): 103.

Asoodeh, M., Bagheripour, P.,2012, Estimation of bubble point pressure from *PVT* data using a power-law committee with intelligent systems, *J. Pet. Sci. and Eng., Vol. 90–91, July 2012, Pages 1-11.*

Bandyopadhyay, P., Sharma, A., 2011, Development of a new semi analytical model for prediction of bubble point pressure of crude oils, *Journal of Petroleum Science and Engineering, Volume 78, Issues 3–4, September 2011, Pages 719-731.*

Becker, P. J., Celse, B., Guillaume, D., Dulot, H., Costa, V., 2015, Hydrotreatment modelling for variety of VGD feedstocks: a continuous lumping approach, Fuel, 139, 133-143.

Coats, K.H., and Smart, G.T., 1986, Application of regression base EOS PVT Program to laboratory data, SPERE, May 277-299.

Dansh, A. Xu, D. H., Todd, A. C., 1992 A grouping method to optimize reservoir description for compositional simulation of gas injection process, SPERE, Aug. 343-348.

De Ghetto, G. and Villa, M. 1994. Reliability Analysis on PVT Correlations. SPE-28904, Presented at the European Petroleum Conference, London, United Kingdom, 25-27 October.

Dindoruk, B. and Christman, P.G. 2001. PVT Properties and Viscosity Correlations for Gulf of Mexico Oils. Presented at the SPE Annual Technical Conference and Exhibition, New Orleans, 30 September-3 October. SPE-71633

Dokla, M.E. and Osman, M.E. 1992. Correlation of PVT Properties for UAE (United Arab Emirates) Crudes. *SPE Form Eval.* (March 1992): 41.

Droham, J.K., 1988, Enhancing the evaluation of PVT data, OSEA paper presented the 7th Offshore South East Asia, Feb. 2-5, Singapore.

Duan, J., Wang, W., Liu, H., Cong, J., 2013, modelling the characterization of the plus fraction by using continuous distribution function, Fluid Phase Equilibria, 345, 1-10.

Elam, F.M. 1957. *Prediction of Bubble Point Pressures and Formation Volume Factors from Field Data.* MS thesis, University of Texas at Austin, Austin, Texas.

Elmabrouk, S., Zekri, A., Shirif E., 2014, The prediction of bubble point pressure and bubble point oil formation volume factor in absence of PVT analysis, PET. Sci. & Tech., 32, 1168-1174.

Elsharkawy, A. M., 2003, An empirical model for estimating the saturation pressures of crude oils, Journal of Petroleum Science and Engineering 38 (2003) 55-77.

Elsharkawy, A.M. and Alikhan, A.A. 1997. Correlations for predicting solution gas/oil ratio, oil formation volume factor, and undersaturated oil compressibility. *J. Pet. Sci. Eng.* 17 (3–4): 291-302.

Emera, M., Samara, H., 2005, Use of genetic algorithm to estimate CO2-oil minimum miscibility pressure- a key parameter in design of CO2 miscible flood, J. Pet. Sci. Eng., Vol. 46, issue 1-2, 37-52.

Farasat, A., Shokrollahi, A., Arabloo, M., Gharagheizi, F., Mohammadi, A., 2013, toward an intelligent approach for determination of saturation pressure of crude oil, Fuel Processing Technology 115 (2013) 201-214.

Frashad, F., LeBlanc, J.L., Garber, J.D. et al. 1996. Empirical PVT Correlations for Colombian Crude Oils. SPE-36105 Presented at the SPE Latin American and Caribbean Petroleum Engineering Conference, Port of Spain, Trinidad and Tobago, 23–26 April. -MS.

Gharbi, R.B.C., Elsharkawy, A.M., 1999, neural network model for estimating the PVT properties of Meddle East crude oils, SPE Reservoir Eval. Eng. 2 (3), 255-265.

Gholami, A., Asoodeh, M., Bagheripour, P., 2014 How committee machine with SVR and ACE estimates bubble point pressure of crudes, *Fluid Phase Equilibria, Volume 382, 25 November 2014, Pages 139-149.*

Glasø, Ø. 1980, Generalized Pressure-Volume-Temperature Correlations. *J Pet Tech.* 32 (5): 785-795.

Heidaryan, E., Moghadasi, J., 2011, Crude saturation pressure through imperial correlation, Thermodynamics 2011, Athens, 31 August- 3 September 2011.

Hasan, M., 1993, PVT Correlations for Indonesian Crude Oils. *Proc.* 22nd Annual Convention of the Indonesian Petroleum Association, Indonesia, Vol.1, IPA93-2.1-070.

Jarrhian, A., Moghadsi, J., Heidaryan, E., 2015, Empirical estimating of black oil bubblepoint (saturation) pressure, J. Petroleum Science and engineering, (126), 69-77.

Jacopy R.H., Berry V.J.,1958, A method for predicting the pressure maintenance for reservoir producing volatile oil, AIME 213, 59-65.

Kartoatmodjo, T., Schmidt, Z. 1994, Large Data Bank Improves Crude Physical Property Correlations. *Oil Gas J.* 92 (27): 51–55.

Kartoatmodjo, T.R.S. and Schmidt, Z. 1991, *New Correlations for Crude Oil Physical Properties*, SPE 23556.

Khairy, M., El-Tayeb, S., Hamdallah, M., 1998. PVT Correlations Developed for Egyptian Crudes. *Oil Gas J.* 96 (18): 114.

Khamekhcki, E., Behvandi, R., Rashid, F., 2011, Prediction of bubble point pressure & asphaltene onset pressure during CO2 injection using ANN & ANFIS models, Pet. Sci. & Tech. Vol. 1 No.2, 35-45.

Labedi, R.M. 1982, *PVT Correlations of the African Crudes*, PhD thesis, CSM.

Labedi, R.M. 1990, Use of Production Data to Estimate the Saturation Pressure, Solution GOR, and Chemical Composition of Reservoir Fluids, SPE-21164-MS. Presented at the SPE Latin America Petroleum Engineering Conference, Rio de Janeiro, Brazil, 14-19 October,

Lasater, J.A. 1958, Bubble Point Pressure Correlations. *J Pet. Tech.*10 (5): 65–67. SPE-957-G.

Levitan, L.L. and Murtha, M. 1999. New Correlations Estimate Pb, FVF. *Oil Gas J.* 97 (10): 70.

Lee, S., 1979, Experimental and theoretical studies on the fluid properties required for simulation of thermal process, Paper SPE 8393.

Li, Y.K., Nghiem, L.X., Siu, A., 1985. Phase behavior computations for reservoir fluids: effect of pseudo-components on the phase diagrams and simulation results. J. Can. Pet. Technol. Nov. – Dec., 29 – 36

Lin, B., Leibovici, C. F., Jorgensen, S. B., 2008, optimal component lumping: problem formulation and solution techniques, Computer & Chem. Eng., 32, 1167-2272.

Macary, S.M. and El-Batanoney, M.H. 1992, Derivation of PVT Correlations for the Gulf of Suez Crude Oils. *Proc. of the* 11th EGPC Pet. Exp. & Prod. Conference, Vol. 1, 374, Cairo, Egypt,

Moharam, H.M., 1995. Prediction of viscosity of heavy petroleum fractions and crude oil using a corresponding states method. Ind. Eng. Chem. Res. 34, 4140 – 4144.

Moradi, B., Malekzadeh, E., Amani, M., Boukadi, F.H., Kharat, R., 2101, Bubblepoint pressure imperial correlation, SPE 132756.

Obomanu, D.A. and Okpobiri, G.A. 1987, Correlating the PVT Properties of Nigerian Crudes. *J. Energy Res. Tech.* 109 (4): 214-217.

Okoduwa, I. G., Ikiensikimama, S.S., 2010, Bubble point pressure correlation for Niger Delta Crude Oils, SPE 136968.

Omar, M.I. and Todd, A.C. 1993. Development of New Modified Black Oil Correlations for Malaysian Crudes. SPE-25338 Presented at the SPE Asia Pacific Oil and Gas Conference, Singapore, 8-10 February.

Owolabi, O.O. 1984. *Reservoir Fluid Properties of Alaskan Crudes*. MS thesis, University of Alaska, Fairbanks, Alaska (May 1984).

Pedersen, K.S., Thomassen, P., Fredenslund, Aa., 1988a. Characterization of gas condensate mixtures. AIChE Spring National Meeting, March 6 – 10, New Orleans, LA.

Peng, D. y., Robinson. D. B., 1976. A new two-constant equation of state , Ind. Eng. Chem. Fund. 15, 59-63.

Petrosky, G.E. Jr. 1990. *PVT Correlations for Gulf of Mexico Crude Oils*. MS thesis, University of Southwestern Louisiana, Lafayette, Louisiana.

Petrosky, G.E. Jr. and Farshad, F. 1998. Pressure-Volume-Temperature Correlations for Gulf of Mexico Crude Oils. *SPE Res Eval & Eng.*1 (5): 416-420. SPE-51395.

Riemens, W.G., Schutle, A.M., Jong, L.N.J., 1988. Birba field PVT variations along the hydrocarbon column and confirmatory field tests. J. Pet. Technol. Jan., 83 – 88

Rosengger, L., Wu., R.1990, Integrated PVT data characterization- lessons from four case histories, J. Can. Per. Tech., 38, (13), (paper 97-05)

Shojaei, M., Bahrami, F., Barati, P., Riahi, S., 2014, Adaptive neuro-fuzzy approach for reservoir oil bubble point pressure estimation, *J. of Natural Gas Sci. and Eng., Volume 20, September 2014, Pages 214-220.*

Soave, G. 1972. Equilibrium constants from a modified Redlich-Kwong equation of state, Chem. Eng. Sci. 27, 1197-1203.

Standing, M.B. 1947. A Pressure-Volume-Temperature Correlation for Mixtures of California Oils and Gases. *API Drilling and Production Practice* (1947): 275-287.

Tale, M., Alquraishi, A., Al Mushaigeh, E. 2009, New model for crude oil saturation pressure prediction using alternation conditional expectation, Offshore Mediterranean Conf. & Exhibit. Italy 25-27, 2009.

Talebi, R., Ghiasi, M. Talebi, H., Mohammadyian, M., Zendehboudi,S., Arabloo, M., Bahadori, A., 2014, Application of soft computing approaches for modelling saturation pressure of reservoir oil, *J. Nat. Gas Sci. and Eng., Vol. 20, Sep. 2014, 8-15.*

Vazquez, M., Beggs, H. D. 1980, Correlations for Fluid Physical Property Prediction. *J Pet. Tech.* 32 (6): 968-970. Vazquez, M.E. 1976. *Correlations for Fluid Physical Property Prediction,* MS thesis, University of Tulsa.

Velarde, J., Blasingame, T.A., and McCain Jr., W.D. 1997. Correlation of Black Oil Properties at Pressures below Bubble Point Pressure - A New Approach. Presented at the Annual Technical Meeting of CIM, Calgary, Alberta, 8–11 June. PETSOC 97-93.

Vogel, J.R., Yarborough, L. 1980, the effect of nitrogen on phase behavior and phase properties of reservoir fluids, SPE 8815.

Whitson, C. H., 1980, Characterizing hydrocarbon plus fraction, SPEJ, 24 (4), 683-694.

Williams, C.A., Zana, E.N., 1980. Use of the Peng-Robinson equation of state to predict hydrocarbon phase behavior and miscibility for fluid displacement. SPE 8817, SPE/DOE on Enhanced Oil Recovery, April 20–23, Tulsa, OK.

Appendix

In this appendix, sample calculations are presented to show the deficiency of some of the published methods.

Input parameters:

N_2	CO_2	H_2S	C_1	C_2	C_3	C_4	C_5	C_6	C_{7+}	GR.C7+	MWC7+	T, °F	Ps, psia
0.65	0.02	0	45.02	12.45	8.93	6.03	3.02	1.44	22.44	0.81	184	140	3002

Ahmadi et al (2014)

Ahmadi et al. proposed an 8 steps calculation (copied from their paper) to estimate the BPP as follows:

$$\mathrm{BPP} = A \times Mw_{C7+} - \frac{B}{C_{7+}} + C \times \mathrm{Vol./Inter.} + D \times SG_{C7+} + \frac{E}{SG_{C7+}} + \frac{F}{Mw_{C7+}} + T \times 505.967 + \frac{2.45789 \times 10^{6}}{T} + \frac{G}{\mathrm{Vol./Inter.}} \quad (5)$$

$$A = 1488.46 - SG_{C7+} \times 1087.64 - \frac{601.512}{SG_{C7+}} + 1670.58 + T \times 0.40771 + \frac{6345.73}{T} + \frac{8.72863}{\mathrm{Vol./Inter.}} \quad (6)$$

$$B = \frac{4132.53}{\mathrm{Vol./Inter.}} + \frac{6.1977 \times 10^{7}}{Mw_{C7+}} + T \times 3181.39 + \frac{1.06393 \times 10^{8}}{T} \quad (7)$$

$$C = \frac{20964.7}{C_{7+}} - \frac{1.76162 \times 10^{6}}{Mw_{C7+}} - T \times 19.1963 - Mw_{C7+} \times 21.5427 + \frac{12851.5}{SG_{C7+}} - \frac{712613}{T} - 3477.49 + SG_{C7+} \times 9532.73 \quad (8)$$

$$D = 1.05345 \times 10^{+06} - \frac{1.50748 \times 10^{8}}{Mw_{C7+}} - T \times 676.048 - \frac{1.20252 \times 10^{7}}{T} - \frac{12178.2}{\mathrm{Vol./Inter.}} \quad (9)$$

$$E = 641357 - \frac{1.02891 \times 10^{8}}{Mw_{C7+}} - T \times 301.034 - \frac{1.91446 \times 10^{6}}{T} - \frac{3873.62}{\mathrm{Vol./Inter.}} \quad (10)$$

$$F = 2.51471 \times 10^{+08} + T \times 25332.5 + \frac{3.47893 \times 10^{8}}{T} - \frac{244573}{\mathrm{Vol./Inter.}} \quad (11)$$

$$G = 20125.6 - T \times 23.9271 - \frac{703193}{T} \quad (12)$$

Ahmadi et al. define the Vol. /Inter. as the mole percentage ratio of the volatile (C_1& N_2) to the intermediate components (C_2 to C_6, CO_2 and H_2S). Application of the above equations from their paper using the input parameters results in the following:

Vol./Inter.	A	B	C	D	E	F	G	BPP
1.43211	1643.9	1545062	-271.127	45122.71	23642.47	2.57E+08	11753	1,794,123

The calculated saturation pressure (BBP) from Eq.5 is 1,794,123 psia which is unreasonable. This indicates a deficiency in the model.

Gholami et al (2014)

Gholami et al. presented 3 steps calculation (copied from their paper) to estimate the BPP as follows:

$$Tr(M) = \sum_{i=0}^{5} B_i M^i \quad (15)$$

Application of Eq. 15 resulted in the following

N_2	0.00584
CO_2	0.060917
H_2S	0.050636
C_1	-0.47658
C_2	-0.41547
C_3	-0.45083
C_4	-0.02003
C_5	0.068769
C_6	0.273921
C_{7+}	1.814523
SG.C7+	-0.0197
MWC7+	37.03525
T., F	-0.11139

Application of Eq. 16

$$Tr(P_b) = Tr(N_2) + Tr(CO_2) + Tr(H_2S) + Tr(C_1) + Tr(C_2) + Tr(C_3) + Tr(C_4) + Tr(C_5) + Tr(C_6) + Tr(C_{7+}) + Tr(SG_{C7+}) + Tr(MW_{C7+}) + Tr(T) \quad (16)$$

Resulted that Tr(Pb) = 37.8159

$$P_b = (57.0840387741 \times (Tr(P_b))^2) + (1166.9102304199 \times (Tr(P_b))) + 2207.3573204358 \quad (17)$$

Application of Eq. 17 resulted in saturation pressure (P_b) = 127,967 psia, which is unreasonable, compared to the measured one is 3002 psia. This indicates that there is a deficiency in the model.

Jarrahian et al (2015)

Jarrahian et al. proposed the following Eq. (copied from their paper) to estimate the saturation pressure:

$$P_{sat} = A_0 \exp\left(A_1 z_{N_2} + A_2 z_{CO_2} + A_3 z_{H_2S} + A_4 z_{C_1}^{A_5} + A_6 \sum_{2}^{6} z_{C_i} + A_7 z_{C_{7+}}^{A_8} \gamma_{C_{7+}}^{A_9} Mw_{C_{7+}}^{A_{10}} + A_{11}(T - 459.67)\right) \quad (2)$$

Application of Eq. 2 resulted in saturation pressure (Ps) = 1812 psia, compared to the measured one is 3002 psia. This indicates that model has inaccuracy problem. This has been the pattern in all the crude oil samples shown in Table 4 and figures 6A through 6D.

Table 1- Summary of attempts to calculate the saturation pressure of crude oils

Method based on production information	Method based on compositional data
Standing, 1947; Elam, 1957; Lasater, 1958; Vasquez, 1976; Vasquez and Beggs, 1980; GlasØ, 1980; Labedi, 1982&1992; Owolabi, 1984; Al-Marhoun, 1985; Obomanu and Okpobiri, 1987; Asgarpour et al., 1989; Al-Najar et al., 1988; Dokla and Osman, 1990; Petrosky, 1990; Kartoatmodjo and Schmidt, 1991&1994; Macary and El-Batanoney, 1992; Omar and Todd, 1993; Hasan, 1993; De Ghetto and Villa, 1994; Farashad et al., 1996; Almehaidaeb, 1997; Elsharkawy, 1997; Velarde et al., 1997; Kairy et al., 1998; Petrosky and Farshad, 1988; Al-shammasi, 2001; Gharbi & Elsharkawy 1999 Levitan and Murtha, 1999; Dindrouk and Christman, 2001 Moradi et al., 2010 Okoduwa and Ikiensikimama 2010 Asoodeh and Bagheripour, 2012 Shojaei et al., 2014 Talebi et al., 2014 Elmabrouk et al., 2014 Kanu and Ikiensikimama, 2014 Jarrahian et al 2015	Elsharkawy , 2003 Tale et al., 2009 AlQuraishi, 2009 Bandyopadhyay and Sharma, 2011 Khamehchi et al., 2011 Farasat et al., 2013 Ahmadi et al., 2014 Gholami et al. 2014 Jarrahian et al 2015

Table 2- Description of the data used to develop and test the extended model

Input	Training model (179 Data)			Testing (28 Data)			Testing (7 new experimental)		
	Min	Ave	Max	Min	Ave	Max	Min	Ave	Max
N_2	0.00	0.35	1.67	0.00	0.27	1.67	0.10	0.26	0.35
CO_2	0.00	1.27	11.37	0.00	1.17	11.37	0.14	0.41	0.83
H_2S	0.00	0.15	3.68	0.00	0.15	3.22	0.00	0.00	0.00
C_1	13.16	33.86	74.18	19.50	34.79	73.36	20.64	28.25	42.79
C_2	3.36	8.07	13.71	4.64	8.26	11.63	6.69	7.88	10.82
C_3	0.89	6.59	11.87	2.46	6.76	12.03	6.05	6.59	7.10
C_4	0.95	4.46	8.4	1.66	4.41	6.58	4.03	5.05	5.93
C_5	0.40	3.21	5.95	1.00	3.11	5.23	3.36	4.33	5.52
C_6	0.00	3.14	6.37	0.99	3.15	5.46	2.88	3.74	4.33
C_{7+}	10.72	38.89	57.73	11.18	37.88	51.22	27.61	43.49	51.31
$SG.C_{7+}$	0.74	0.87	0.959	0.70	0.86	0.93	0.85	0.89	0.92
MWC_{7+}	134	240	368.9	105	223	292	216	262	298
T, ° F	58	156	319	58	165	261	121	157	230
Ps, psia	537	2198	5065	1025	2307	5000	1227	1779	3344

Table 3: Error analysis of the proposed models in comparison with all the investigated methods

Method	ARE	AARE	SD	R^2%
PR-EOS	6.24%	10.19%	1028	97.9
SRK-EOS	-3.12%	10.36%	1024	96.7
Elsharkawy (2003)	0.85%	8.57%	1015	96.4
Alquraishi (2009)	-6.06%	10.16%	1088	96.7
Bandyopadhyay (2011)	8.05%	12.28%	1054	96.8
Farasat (2013)	NA	NA	NA	NA
Gholami et al. (2014)	?	?	?	?
Ahmadi et al. (2014)	?	?	?	?
Jarrahian et al (2015)	35.52	35.56	983	95.2
New Extended Model	-0.37%	8.20%	1030	97.6
New Lumped Model	-0.50%	8.46%	1036	97.4

ARE = Average Relative Error

AARE = Average Absolute Relative Error

SD = Standard Deviation

R^2 = Correlation Coefficient

NA = Calculation was not possible, because the weight and bias were not provided.

? = The method produces unreasonable results as indicated in the appendix.

Table 4- Comparing model accuracy for calculations of saturation pressures for various crude oils

Volatile-Oil-Samples (C7+<30)																						
Fluid Source	index	N2	CO2	H2S	C1	C2	C3	C4	C5	C6	C7+	GR.C7+	MWC7+	T, °F	Ps, psi	SRK	PR	Elsh	Alq	Bandy	Jarr	Extd	Lump
Rosenegger	7	0.03	8.39	0	47.4	10.3	6.12	4.2	2.89	2.05	18.6	0.83	180	295	4000	3933	3879	4460	4420	4165	3317	4139	4178
Vogel-80	13	0.65	0.02	0	45	12.5	8.93	6.03	3.02	1.44	22.4	0.81	184	140	3002	2624	2548	2742	2895	2643	1812	3019	2996
Dansh-92	18	0	0	0	46.8	8.77	7.44	4.01	2.56	4.02	26.4	0.77	158	212	2941	2550	2594	3248	3617	3269	1944	3244	3374
	31	0.44	0.38	0	49.1	7.6	6.13	3.84	2.52	2	28	0.84	231	199	3739	3522	3507	3418	3951	3508	2309	3625	3644
Middle East	45	0.03	0.97	0	41.6	10.8	7.52	4.5	3	2.43	29.1	0.85	208	241	3069	3109	2979	3080	3513	3138	1962	3127	3140
Coats-86	51	0.55	1.03	0	36.5	9.93	8.85	6	3.78	3.56	30.4	0.84	200	234	2746	2537	2472	2714	2945	2643	1682	2777	2720

Black-Oil -Samples (35<C7+<55)																						
Fluid Source	index	N2	CO2	H2S	C1	C2	C3	C4	C5	C6	C7+	GR.C7+	MWC7+	T, °F	Ps, psi	SRK	PR	Elsh	Alq	Bandy	Jarr	Extd	Lump
Coats-86	73	1.64	0.08	0	28.4	7.16	10.5	8.4	3.82	4.05	36	0.84	252	131	1694	1674	1618	1615	1738	1644	1084	1713	1713
Middle East	88	0.08	1.37	0.69	36	8.67	5.94	2.97	1.5	3.1	39.7	0.9	274	168	2505	2710	2318	2247	2608	2321	1476	2441	2445
Middle East	114	0.29	0.48	0	28.4	8.29	7.38	5.06	3.42	4.41	42.3	0.88	252	133	1632	1672	1477	1544	1748	1517	984	1587	1598
Middle East	122	0	1.25	0	33.4	9.24	6.07	2.4	1.62	2.78	43.3	0.85	252	230	2110	2313	2278	2424	2737	2438	1410	2388	2412
Middle East	134	0	0.17	0	28.6	8.63	7.43	3.81	2.61	4.11	44.7	0.86	249	134	1548	1551	1429	1477	1764	1433	937	1503	1567
Middle East	146	0.44	0.83	0	27.8	7.54	6.9	4.61	3.1	3.11	45.7	0.9	272	133	1565	1799	1505	1558	1730	1528	1002	1602	1590
moharam-96	149	0.21	0.34	0	20	7.93	8	6.6	5.87	5.08	45.9	0.86	230	235	900	1353	1290	1531	1514	1225	773	1303	1294
Middle East	162	0.35	0.47	0	26.5	7.71	6.05	4.03	4.08	4.22	46.6	0.89	253	135	1500	1613	1380	1427	1631	1391	923	1478	1456
Middle East	176	0.12	0.51	0	28.8	7.91	6.46	3.17	2.07	3.25	47.7	0.88	250	134	1615	1665	1476	1574	1783	1502	981	1571	1614
Middle East	194	0.21	0.15	0	27.8	7.68	6.19	3.41	1.81	2.84	49.9	0.86	228	134	1590	1532	1377	1545	1680	1426	923	1466	1493
Rosenegger	207	0.37	0.02	0	17.3	6.11	8.23	5.47	3.76	3.96	54.8	0.85	242	156	537	938	888	914	1064	836	558	661	731

Newly-Measured Kuwait University Laboratory

Fluid Source	index	N2	CO2	H2S	C1	C2	C3	C4	C5	C6	C7+	GR.C7+	MWC7+	T, °F	Ps, psi	SRK	PR	Elsh	Alq	Bandy	Jarr	Extd	Lump
KU-1	209	0.33	0.22	0	25.6	6.87	6.39	5.61	4.68	4.33	46	0.879	222	133	1595	1360	1530	1442	1514	1394	860	1357	1337
KU-2	210	0.35	0.47	0	26.5	7.71	6.05	4.03	4.08	4.22	46.6	0.893	253	135	1500	1400	1620	1427	1631	1391	924	1478	1456
KU-3	211	0.27	0.74	0	23.1	7.2	6.45	5.93	4.66	4.16	47.5	0.906	290	138	1283	1140	1360	1284	1448	1296	800	1287	1219
KU-4	212	0.22	0.26	0	20.6	6.69	6.65	5.11	5.52	3.6	51.3	0.919	298	171	1227	1120	1300	1188	1415	1170	742	1176	1131
KU-5	213	0.19	0.14	0	27.8	8.37	6.88	5.28	4.33	3.85	43.2	0.89	278	121	1524	1300	1560	1402	1659	1409	928	1534	1489
KU-6	214	0.33	0.24	0	31.3	7.48	6.58	4.86	3.7	3.17	42.3	0.9	274	171	1978	1800	2200	1975	2242	1985	1224	2060	2026
KU-7	215	0.1	0.83	0	42.8	10.8	7.1	4.51	3.36	2.88	27.6	0.85	216	230	3344	3070	3400	3083	3568	3139	2003	3213	3200

Low-Temperature Oil Samples (59°F<T<71°F)

Fluid Source	index	N2	CO2	H2S	C1	C2	C3	C4	C5	C6	C7+	GR.C7+	MWC7+	T, °F	Ps, psi	SRK	PR	Elsh	Alq	Bandy	Jarr	Extd	Lump
Middle East	101	0	1.12	0.21	27	10.8	9.23	4.97	2.62	3.33	40.8	0.88	249	71	1247	1090	1033	1040	1235	564	808	1148	1205
Middle East	117	0.36	0.17	0	27.2	8.93	8.6	6.07	3.71	2.55	42.4	0.88	271	68	1130	1028	1245	1094	1248	766	820	1239	1228
Middle East	135	0	0.17	0	28.6	8.63	7.43	3.81	2.61	4.11	44.7	0.86	249	70	1270	1055	1255	1130	1311	783	826	1229	1297
Middle East	138	0.14	0.54	0	29.4	7.96	6.19	3.96	3.25	3.84	44.7	0.9	290	59	1265	1065	1350	1191	1301	880	894	1408	1390
Middle East	147	0.44	0.83	0	27.8	7.54	6.9	4.61	3.1	3.11	45.7	0.9	272	67	1320	1125	1400	1201	1268	921	879	1319	1312
Middle East	173	0.06	0.81	0	31.3	7.42	5.68	2.82	1.63	2.64	47.6	0.89	270	68	1460	1215	1535	1441	1463	1026	975	1548	1576
Middle East	193	0.16	0.3	0	24.7	7.51	6.92	4.76	2.65	3.43	49.6	0.88	239	71	1050	975	1156	1008	1107	701	714	958	986

Elsh: Elsharkawy (2003)
Extd: Extended model
Alq: AlQuraishi (2009)
Lump: Lumped model
Band: Bandyopadhyay and Sharma (2011)
Jarr: Jarrahian et. al (2015)

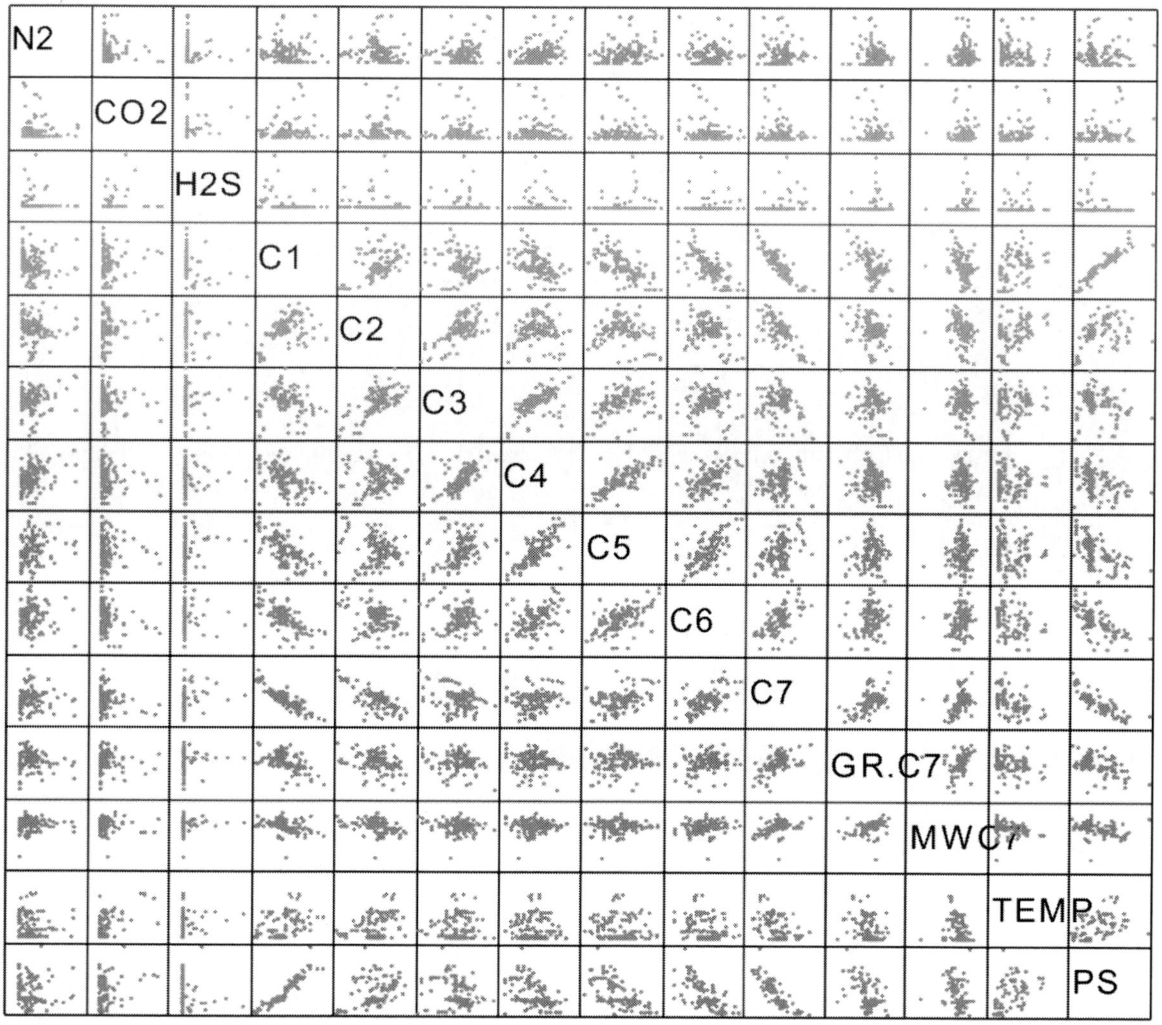

Figure 1: Correlation matrix for all the variables in the oil data bank

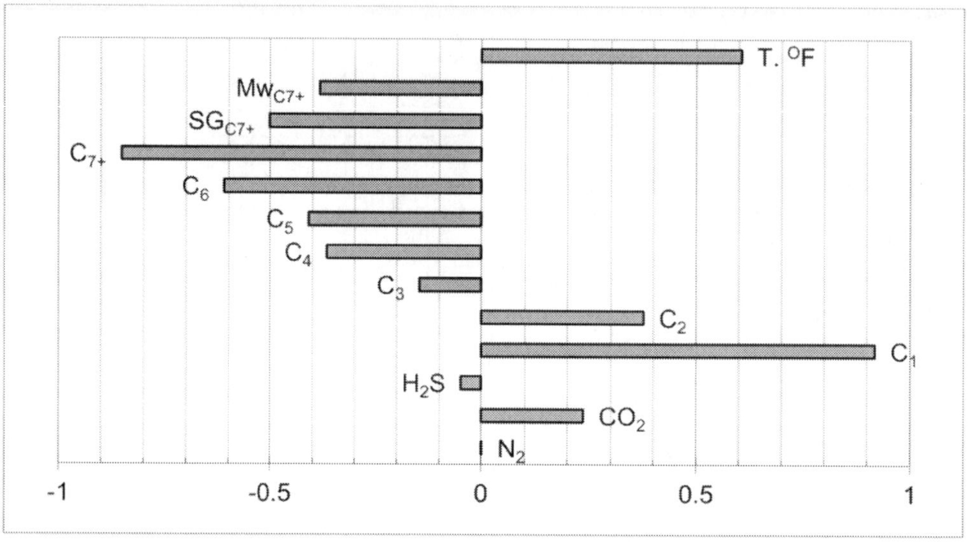

Figure 2: Correlation coefficient for all the variables in the training model (179 data)

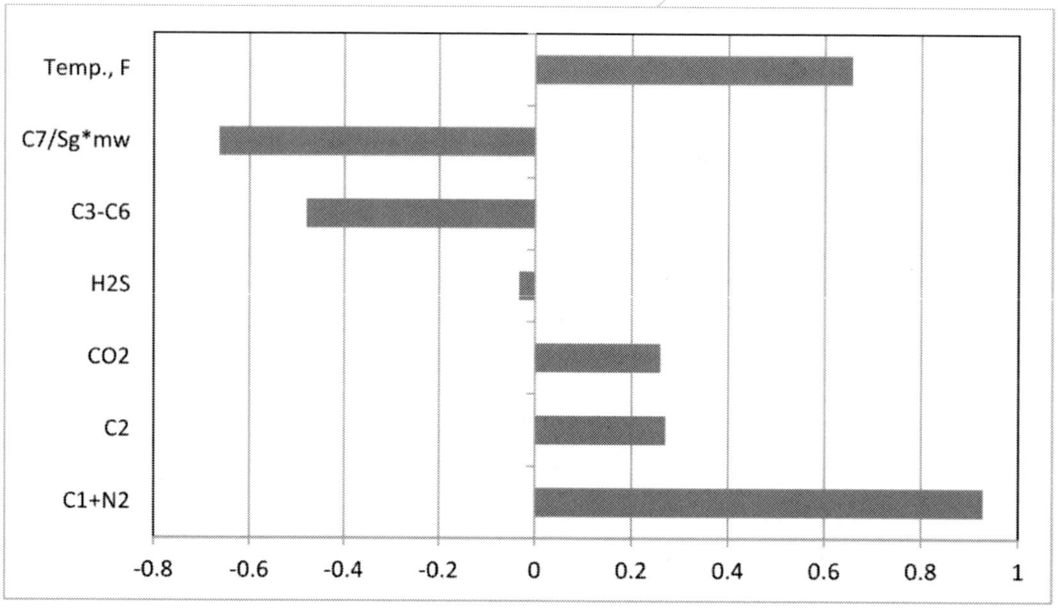

Figure 3: Correlation matrix for the lumped model (179 data)

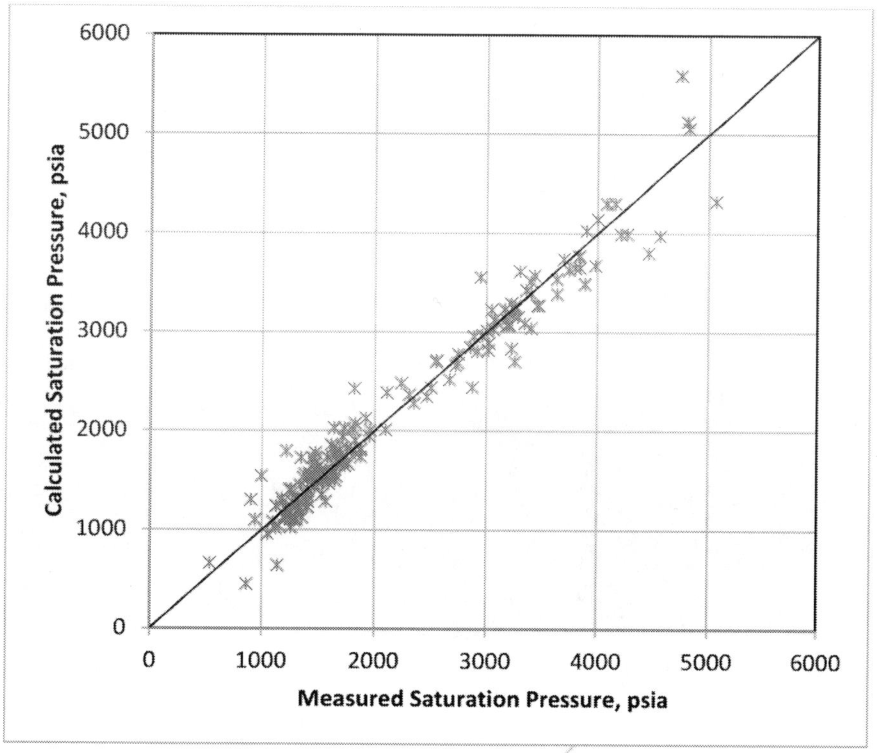

Figure 4A: Cross plot of the extended model using 179 training data

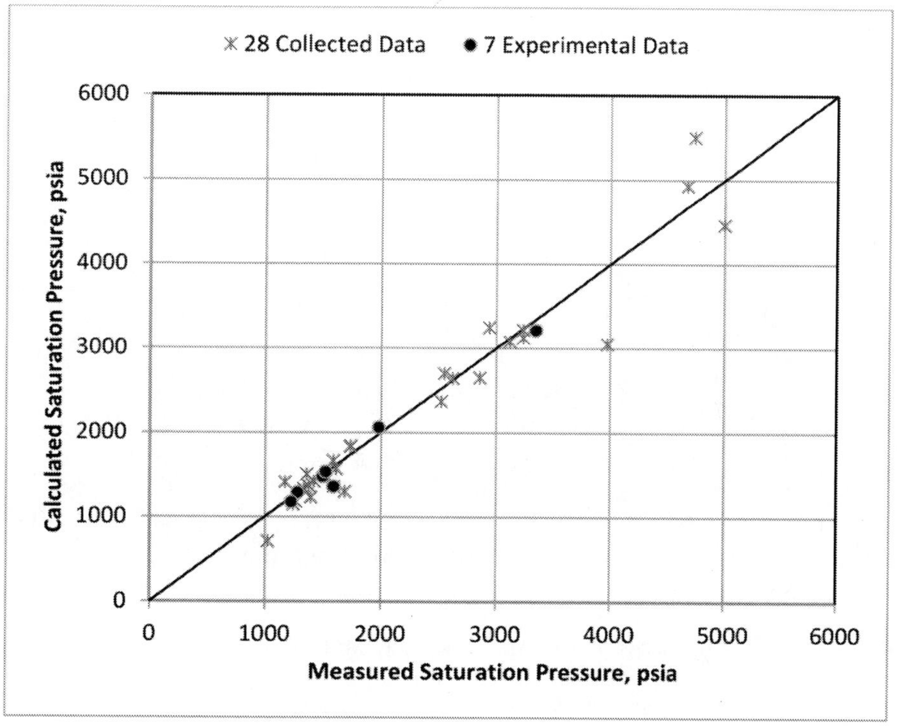

Figure 4B: Cross-plot of the extended model using 35-testing data

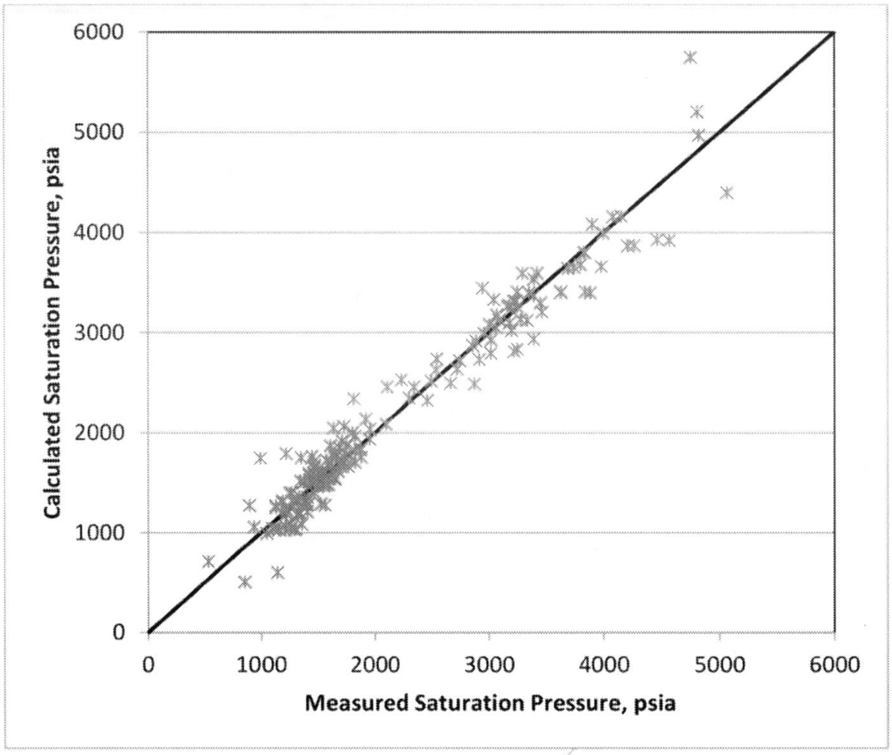

Figure 5A: Cross-plot of the lumped model using 179 training data

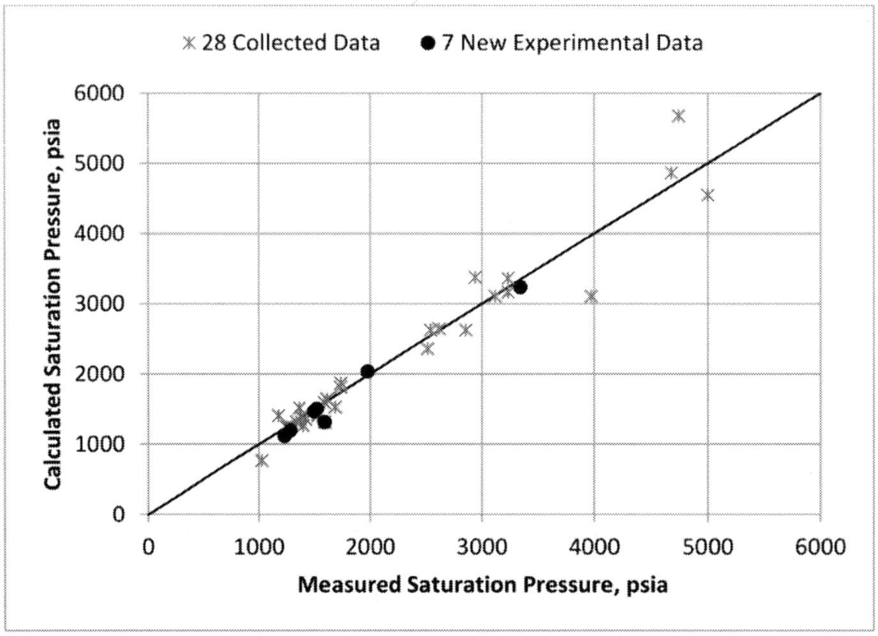

Figure 5B: Cross-plot of testing data for the lumped model

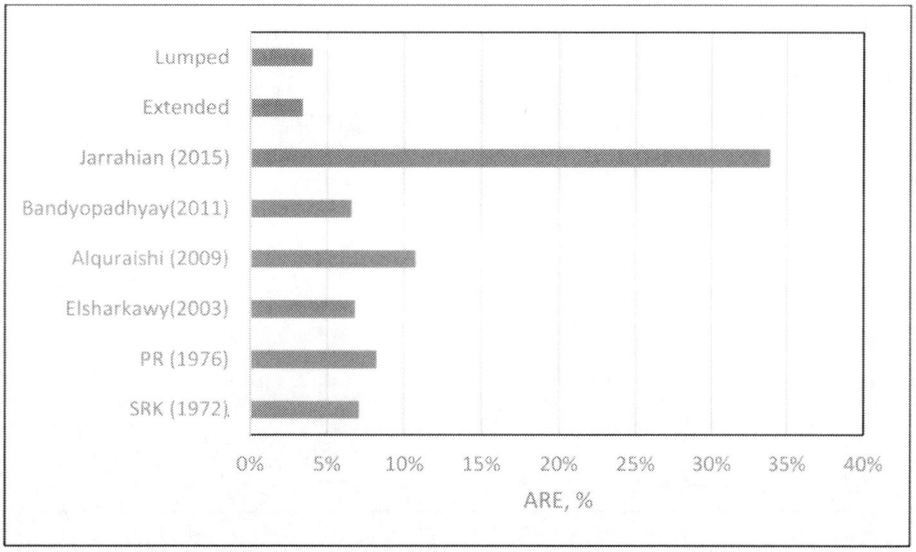

Figure 6A: Comparison of all the models' accuracy for the volatile oil samples

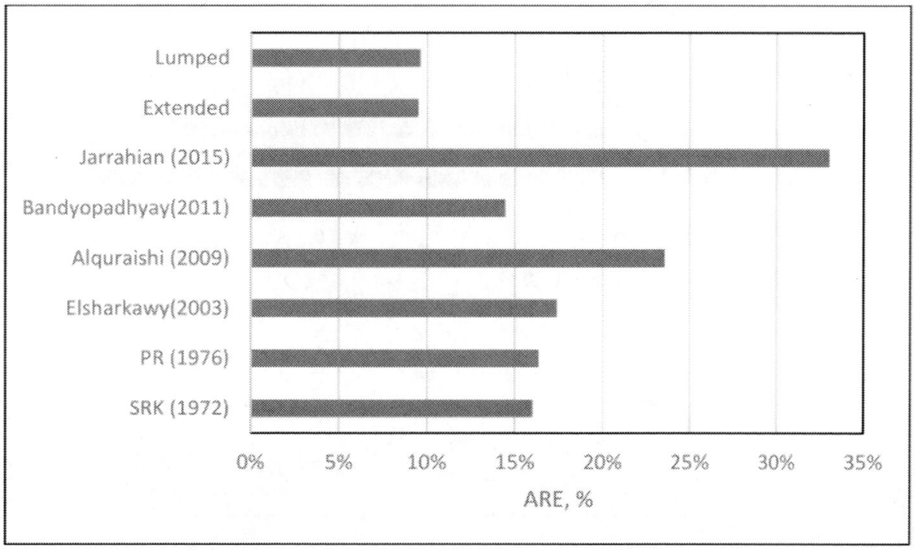

Figure 6B: Comparison of all the models' accuracy for the black oil samples

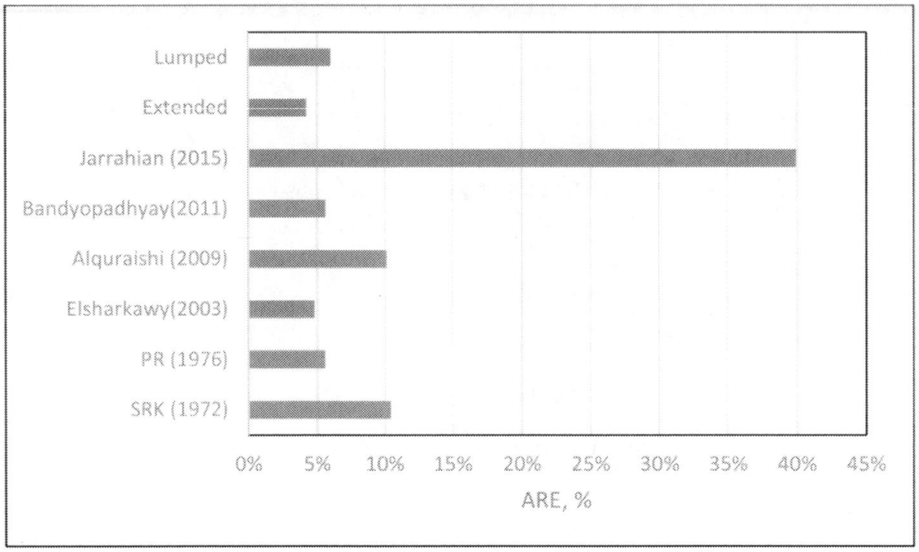

Figure 6C: Comparison of all the models' accuracy for the newly measured samples at KU

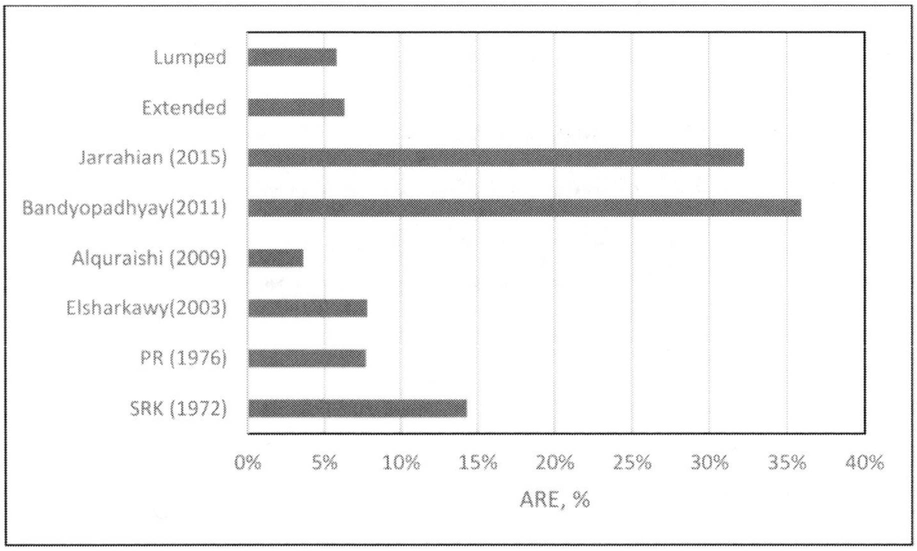

Figure 6D: Comparison of all the models' accuracy for low-temperature oil samples

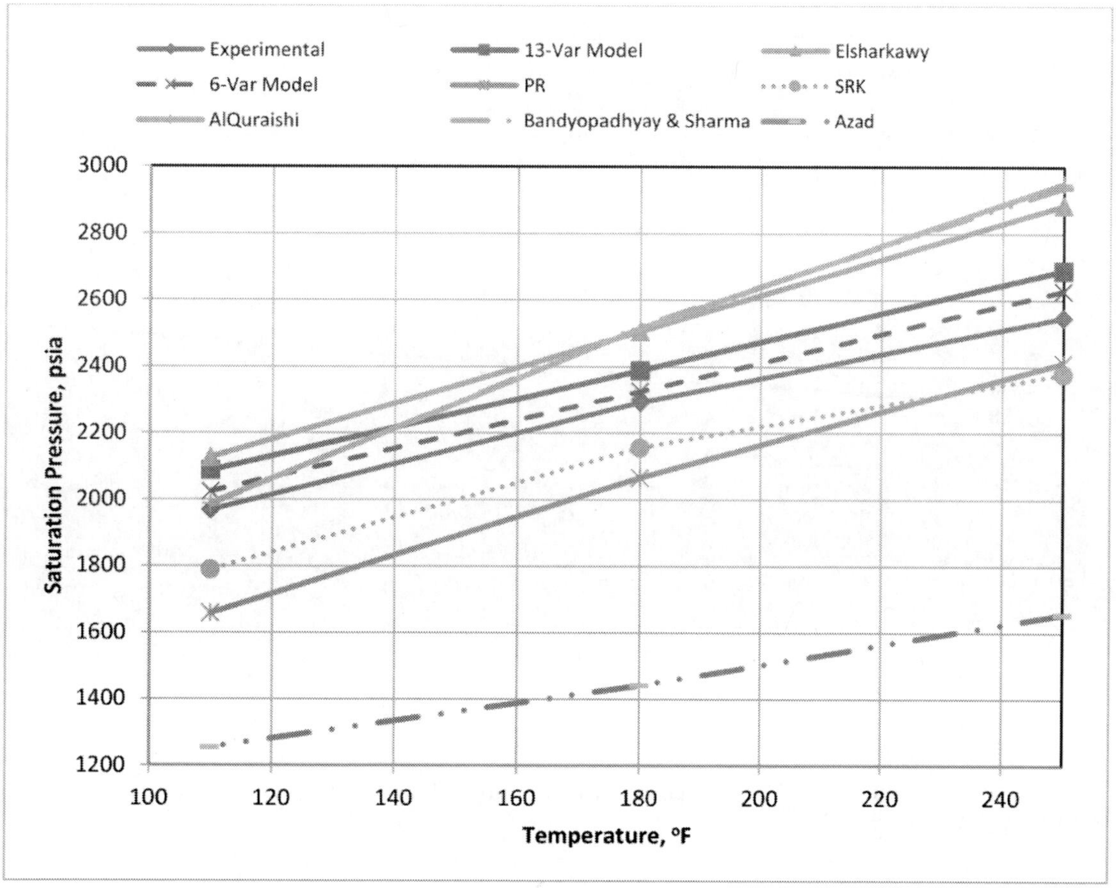

Figure 7: Changes in saturation pressure as a function of temperature for the different models (Sample #70)

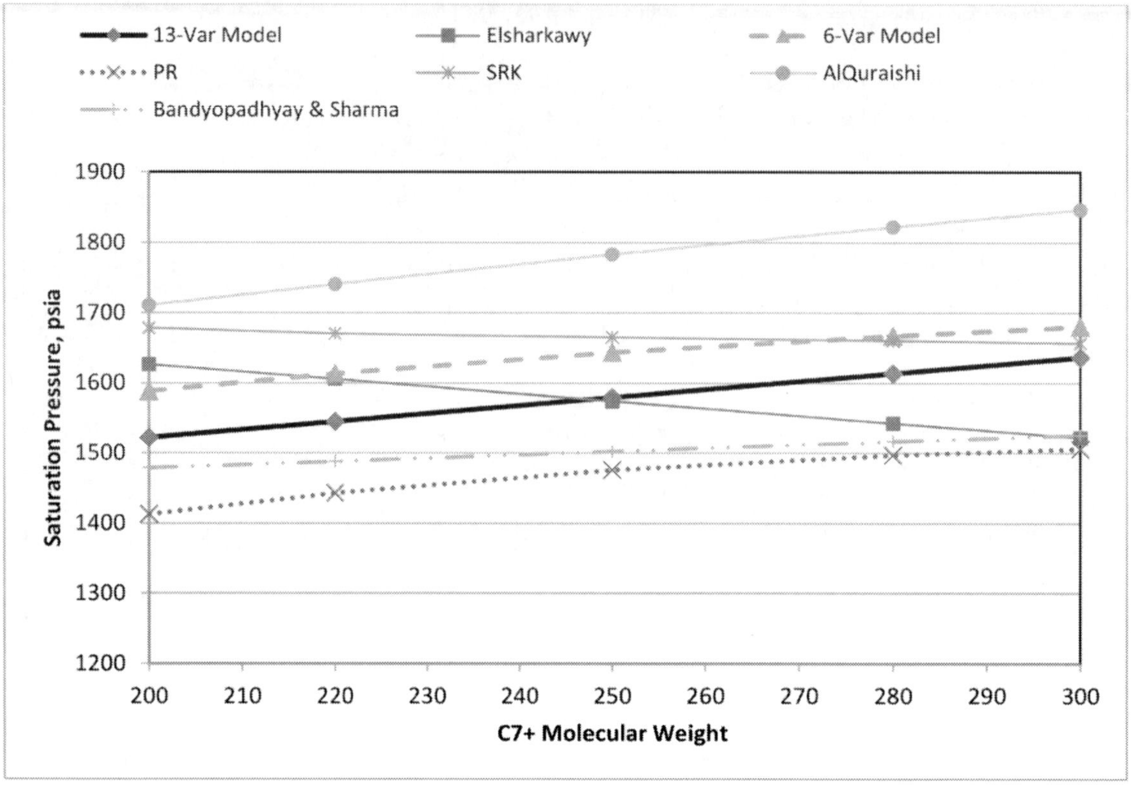

Figure 8: Comparing model performances: Changes in BPP as a function of C_{7+} molecular weight at 134°F (Sample #176)

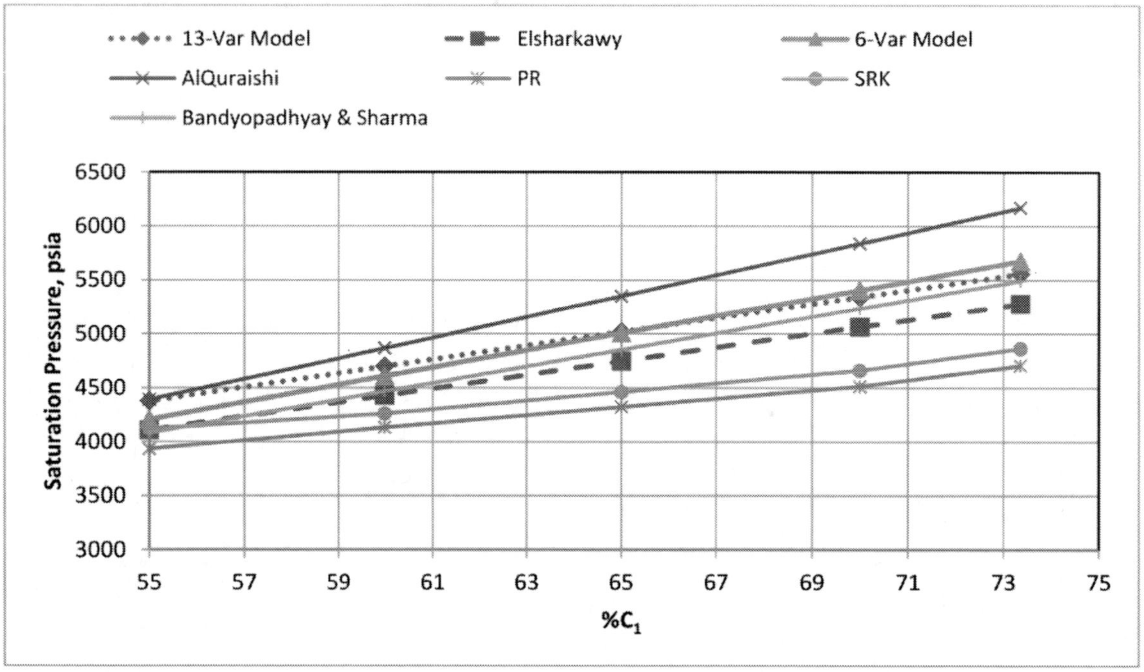

Figure 9: Sensitivity of saturation pressure to methane content at 212°F (sample #2)

444834 Stability and Rheology of Emulsified Acid in Waste Oil: Effect of Emulsifier Concentration, Mixing Speed, Temperature and Droplet Size

Abstract:

Acid in oil emulsion is used in the industry for stimulation purposes. In particular, HCl as the acid phase and diesel as the oil. Emulsifying HCl has numerous benefits over regular HCl. Perhaps the main reason is due to deeper penetration. Apart from that, less corrosion damages are caused since the external phase is hydrocarbon. Several studies showed the success of replacing diesel as an external phase with other hydrocarbon oil, such as crude oil and xylene.

This work utilizes the extra hydrocarbon left unused –or sometimes dumped- from refineries. This unused hydrocarbon is referred to as waste oil. First the chemical composition of waste oil was studied using HPLC, SARA fractionation as well as FTIR. This helped in selecting the proper emulsifier agent for the waste oil. The HCl-waste-oil emulsion was prepared using 15wt% HCl with a ratio of 70:30 of acid-to-oil. Extensive work was done to ensure reproducibility of data. This study was done to experimentally investigate the effects of distinct important variables on thermal stability and apparent viscosity of the HCl-waste-oil emulsion. These variables are; emulsifier concentration, mixing speed and the addition of corrosion inhibitor. All experiments were conducted at a high temperature 120°C. To assess the quality of the HCl-waste-oil emulsion, droplet size was measured as a function of temperature (up to 100 °C).

The results showed that the HCl-waste-oil emulsion is a shear-thinning fluid. Power-law model was applicable seamlessly to all of the apparent viscosity data for all of measured temperatures. The optimum conditions were found to be; 0.5vol% and 600 rpm for emulsifier concentration and mixing speed, respectively. Apart from that, the emulsions prepared with corrosion inhibitor were less stable compared to the ones without it. However, the addition of corrosion inhibitor tended to generate monodispersed emulsion but bigger size droplets for temperatures 25°C and 50°C and polydispersed droplets for higher temperatures. Droplet size measurements showed an average size of 1.47 to 3.09 μm. Overall, the lab results showed a promising potential for the HCl-waste-oil emulsion to be used in the field.

This work takes into account; reducing the budget of acidizing job and at the same time environmental concerns. It makes use of the unwanted hydrocarbon from refiners. And, it saves the environment by using the waste oil instead of dumping it.

443836 Oxidative Dehydrogenation of n-Butane on Carbon Catalysts

Abstract:

Butenes and butadiene, which are one of the important industrial chemical products to produce synthetic rubber , are sythesized traditionally by oxidative dehydrogenation (ODH) of n-butane over various mixed metal oxides. Metal-free carbon nanomaterials has recently received much attention by researchers as a catalyst for the ODH reaction due to its environmental acceptability, corrosion resistance, and unique surface properties. For the reason, the reaction system was modelled in this study for the sythesis of butene and butadiene. For the modelling of the reaction system, the catalytic performance of several carbon catalysts, such as graphite and activated carbon, were initially tested. Then, the reaction mechanisms and kinetics were developed based on experimental results. Lastly, the mathematical modeling of the reaction system was optimized to maximize an overall yield of the main products. This approach allowed us to develop the optimized reaction system more efficiently.

This research was respectfully supported by Engineering Development Research Center (EDRC) funded by the Ministry of Trade, Industry & Energy (MOTIE). (No.N0000990)

443837 Blowdown Dynamic Modeling and Analysis of Pressure and Temperature

Abstract:

Vessel blowdown is one of the safety operation procedures that discharges flammable gas or liquid to flare system during emergency situation such as fire. When blowdown occurs, the liquefaction and freezing effect cause a pressurized vessel to be brittle, exposing it to possible hazard situation. Also the blowdown systems are recommended to be accordance with API(American Petroleum Institute) 521 code. API 521 requires pressure in vessel should be half of the initial pressure or be lower than 100 psig in 15 minutes after commencing blowdown in fire case. Therefore, accurate estimation over the change of temperature and pressure of the vessel during blowdown period is crucial for the design of blowdown system. For the reason, researchers have developed a few dynamic blowdown models so far. However, the model was found to be rather inaccurate under certain conditions. Therefore a new dynamic blowdown model was developed in this work. For the accurate analysis, the model algorithm considers both vapour and liquid phase in non-equilibrium condition that the temperatures of vapour and liquid phases are different. Futhermore, heat transfer from the vessel wall to discharge gas at either laminal or turbulent flow in the vessel was considered in the model to increase the accuracy of blowdown estimation. For thermodynamics, both Peng-Robinson and Soave-Redlich-Kwong equations of state were applied to the model, and the results were compared each other. Finally, the simulation results were compared with Haque et al's experimental and steady-state simulation results.

444364 Molecular Simulation of Controlling Ions and Their Role in Low Salinity IOR for Carbonate Reservoirs

Abstract:

Carbonate geology represents around 50% of hydrocarbon-bearing reservoirs in the world and more than 70% in some areas such as the Middle East. Water flooding has not been very efficient in carbonate reservoirs due mainly to the low water wetness. Modifying the ionic composition and salinity of the injected seawater demonstrated that the oil recovery from these carbonate reservoirs could be improved. Although the underlying mechanisms are still a subject of debate, the improved oil recovery has been mostly attributed to the increased water-wetness of the rock through the interactions of specific ions like Ca^{2+}, Mg^{2+}, SO_4^{2-}, and Cl^- , which are present in seawater with the rock surface under favorable conditions

The study of hydrated gas-phase metal ions provides a link between intrinsic chemistry of the isolated ion and its chemistry in solution. The structures and binding enthalpies of a variety of gas-phase and solution mono- and divalent ion hydrates containing up to 12 water molecules have been studied computationally. A DFT and molecular dynamics simulations were employed to study hydrated structures of the form $[\mathbf{M}(H_2O)_n]^{2+}.mH_2O$ for calcium and magnesium and $[\mathbf{M}(H_2O)_n]^{+}.mH_2O$ for sodium, in which there are n water molecules in the first coordination shell and m water molecules in the second coordination shell. The hydration, dehydration, and interactions of metal ions with Cl^{-1} and SO^{-2}_4 have been studied to better understand the mechanism of low salinity IOR. Density functional (DFT) theory using the B3LYP functional, and 6-311++g** basis set has been used. The charge of the metal ion decreases monotonically as the number of water molecules increases in the complex. Extensive calculations of the energetic contribution of individual structural characteristics as well as the energetic trends in binding energy are developed, which will lead to better understanding of how low salinity water IOR works.

Keywords: Hydration, carbonate reservoir, DFT, binding energy, Low Salinity

447119 Development R&D of Manoir Industries and Emphasis on Coating

Abstract:

The Senior VP Technical Support & Projects will present the development R&D of Manoir Industries and especially the coating (it is our lattest development) as well the services we can provide such as metallurgy expertise, advice, convector cleaning, site welding, welding supervision in order to increase the furnace productivity. So it is 60% R&D and 40% services. Manoir Industries manufactures cast components for the petrochemical market and develops high temperature alloys such as the Manaurites.

442925 Identifying Gas Liquid Contactor Performances for Optimized Natural Gas Desacidification Amines Units - from Process Calculations to Experimental Strategy

Abstract:

Amine units are widely used for gas sweetening. The absorber performances of these reactive absorption processes are typically mass transfer limited. The proper selection of gas liquid contactor is thus critical to optimize the size of towers, the solvent flowrate and the duty of the reboiler. This work proposes an approach to identify the most suitable packings for a given application.

The case-study selected here to illustrate the approach deals specifically with a CO_2removal application where the sour gas is at 7 % vol, and the sweet gas specification is 2% vol. The gas flowrate is fixed at 0.18 MMSCFD with a solvent flowrate fixed at the minimum of the thermodynamic constraints (via equilibrium curves). An in-house rate-based simulator with accessible forms of mass transfer correlations, based on a double film model, is used to calculate the absorber height and diameter which are needed to reach the sweet gas specifications. A reference structured packing is compared with standard valve trays.

The calculations show that the bed height required with the first contactor is considerably higher than with the second one while the column diameter is reduced. In this case the use of standard valve trays is more attractive for cost optimization. Interestingly, because mass transfer is accelerated, the absorption rate with the structured packing is found to be thermodynamically limited due to a more severe temperature increase which reduces the driving force in the bottom section of the absorber.

The study determined that the optimum design for the investigated case is a compromise between a full trayed tower, that generally imposes a larger diameter, and a packed tower where the higher mass transfer forces result in a temperature rise in the bottom of the tower which in turn imposes a height increase and a much higher solvent flowrate. . To determine the mass transfer characteristics of a gas-liquid contactor providing this optimum, a sensitivity analysis has been conducted. Mass transfer parameters, i.e., liquid and gas side mass transfer coefficients and interfacial area, were varied by a factor 2 relative to the ones obtained with the reference structured packing. Calculations show that one should use a contactor that reduces liquid phase resistance, specifically, the mass transfer conductance $k_L a_e$ should be increased by at least 50%. This target could be reduced to +30% if one could simultaneously reduce the gas phase conductance $k_G a_e$ by 20%. This result can be explained by the link between mass and heat transfer (Chilton-Colburn analogy) whereby reducing $k_G a_e$decreases the temperature level which in turn thermodynamically increases the mass transfer driving force.

The next step will be to review the literature more extensively to make a short list of pre-selected random and structured packings that best meet the targets. Experimental tests will then be run on IFPEN facilities to measure flooding limits and mass transfer parameters in a small 150 mm diameter column. This step is required to confirm packing's behavior, considering the disparity of the values reported by different authors in the literature. The last step is to consolidate the existing correlations for the most promising packing, and mitigate differences between various authors by performing characterization tests on larger diameter test columns. The final correlations will be implemented in our in-house rate based simulator for industrial use.

The present approach can be applied to determine the required mass transfer characteristics for optimizing other gas treatment applications. However, the conclusions reached here as regards to mass transfer targets and the problematic temperature levels, are valid only to the present case –study, and are not directly applicable to other gas sweetening applications (selectivity, LNG, …). For example, in a case where H_2S and CO_2 are treated simultaneously the mass transfer limitations could be very different.

443784 Measurement and Correlation of the Mutual Solubility of Diisopropylamine + Water and Triethylamine + Water Systems at High Pressure

Abstract:

The mutual solubility data of diisopropylamine + water and triethylamine + water systems were measured at the pressure up to 100 MPa and the temperature range was 280.2~316.2 K. Experiments were performed in a high-pressure autoclave with a moveable piston. The experimental data were correlated satisfactorily with NRTL activity coefficient model and the binary parameters were obtained.

444161 Novel Microwave Synthesis Carbon Dots from PINE Bark and Its Application for Heavy Metal Sensing

Monday, April 11, 2016
Exhibit Hall E (George R. Brown)

Saheed Sanni, Chemistry, Vaal University of Technology, Vanderbijlpark, South Africa, Puthalapattu ReddyPrasad, Chemistry, Vaal University of Technology, Vanderbiljpark, South Africa and Augustine E. Ofomaja, Chemistry, Vaal university of Technology, Vanderbiljpark, WV, South Africa

NOVEL MICROWAVE SYNTHESIS CARBON DOTS FROM PINE BARK AND ITS APPLICATION FOR HEAVY METAL SENSING

Sanni S.O., ReddyPrasad P., Ofomaja A.E.

Biosorption and Waste water Treatment Research Laboratory, *Department of Chemistry, Faculty of Applied and Computer Sciences, Vaal University of Technology, P. Bag X021, Vanderbijlpark-1900, South Africa.*

INTRODUCTION

Microwave pyrolysis for the production of carbon has the advantages of being cheap, produces high yield of carbon and is absolutely a green synthetic method [1]. In recent times, the use of microwave pyrolysis have been extended to the preparation of carbon dots. Carbon nanodots (C-dots) falls under the class of carbon nanoparticles with size below 10 nm[2]. C-dots are carbon materials which processes high water solubility and strong fluorescence depending on their size, edge, shape, surface ligands, and defects. Since C-dots preparation can be achieved from difference carbonaceous sources and by varying a range of parameters during their synthesis, the properties of the final product may vary depending on these factors. C-dots have sp^2 characteristics and possess several functional groups on their surfaces that makes them highly hydrophilic and easily functionalized with various organic, polymeric, inorganic, or biological spices[3]. Several synthesis methods that have been explored includes electrochemical synthesis, combustion/thermal/hydrothermal/acidic oxidation, supported synthesis, microwave/ultrasonic, arc discharge, laser ablation/passivation, and plasma treatment[4]. C-dots processes excellent such as water dispersibility, chemical stability and photostability, ease of modification, excitation-dependent multicolor emission, low toxicity, and good cell permeability, great efforts have been paid to their potential applications in biosensing, bioimaging, drug delivery, and other biological related aspects.

METHODS

The carbon dots (C-dots) with green luminescence was synthesized from microwave assisted pyrolysis of pine bark waste our carbon source at 800, 1000 1200 W for 1 Hr. The dots produced were characterized using FTIR, TGA, XRD, UV-Vis and Raman spectroscopy. The dots were applied for heavy metal sensing.

RESULTS AND DISCUSSION

The functional groups present on the C-dots (**Fig.1**) includes -OH stretching band mode at 3413 cm^{-1}, C-H stretching at 2934 and 2845 cm^{-1}, stretching peaks of C=O, C=C at 1731, 1604 and 1437 cm^{-1}. The XRD pattern (**Fig 2a &b**) of produced C-dots depicts sharp peak at 20.9^{o} and 26.8^{o}, which is highly crystalline cores and possess amorphous nature of C-dots. This peculiar peak is ascribed to 002 of the graphitic carbon, which in line with inter-planar spacing of 3.77 Å with higher spacing than bulk graphite at 3.34 Å. The thermogravimetric and differential thermal analysis display (**Fig 3**) that the C-dots possess thermal stability more than 100 ^{o}C and mass losses occurs afterwards. The thermal analysis depicts 3 endothermic peaks at 331, 542 and 735 ^{o}C, which are ascribed to oxygenated groups such as carboxyl, hydroxyl, carbonyl and phenols. The optical properties of the synthesize C-dots were explored using UV-vis absorption and photoluminescence (PL) spectroscopy. The UV-vis spectra display (**Fig 4**)intense absorption peak in the range of 260 to 290 nm, which is ascribed to π-π^* transition of the aromatic sp2 domain and also presence of multiple polyaromatic chromophores. A shoulder peak at 330 nm which is ascribed to n-π^* transitions of C=O and structural defects of the C-dots. PL spectrum of the C-dots (**Fig 5a**)exhibits an intense maximum emission peak at 432 nm upon excitation at 330 nm, which also shows characteristic yellowish green color under sunlight and intense blue coloration UV lamp (365 nm). The PL properties of as prepared C-dots was varied at (**Fig5b**) different excitation wavelength from 300 to 405 nm. There was red-shifting in emission peak as the excitation peak was increased from 300 to 405 nm and also simultaneously decrease in fluorescent intensity, which is ascribed to quantum effect and surface defects. This behavior of the prepared C-dots is dependent on excitation-fluorescence, which is accordance with other report [13]. The quantum yield of synthesized C-dots was measured to be 17% at maximum excitation at 330 nm

Fig.6 shows the Raman spectrum of the PB-GQDs, which has been usually used to confirm the quality of the prepared PB-GQDs. There is two major Raman peaks at 1370 cm^{-1} (D band), 1580 cm-1 (G band). The peak of G band corresponds to the Sp^2 bonded carbon atoms in a two-dimensional (2D) hexagonal lattice and the peak of G band is associated with the vibrations of carbon atoms with dangling bonds in the termination plane of tangled graphite or glassy carbon. The D/G ratio band ratio is ID/IG: 0.68 is confirm the formation of graphene like structure from synthesized pine bark.

Fluorescence detection of Cu^{2+} ion

The detection of Cu^{2+} was performed at room temperature in phosphate buffer (PBS) solution pH 4.3. 1.0 mL of PB-CQDs solution was added into 0.5 mL of PBS buffer, followed by the addition of 1.0 mL different concentrations of Cu^{2+}. The fluorescence emission spectra were recorded after reaction for 5 min at room temperature (~28 °C).

Selectivity and sensitivity studies

Moreover sensitivity and selectivity is additional significant parameter to assess the performance of the sensing system. Consequently, we examined the fluorescence intensity changes in the presence of representative metal ions such as including Cu^{2+}, Hg^{2+}, Ca^{2+}, Al^{3+}, Ba^{2+}, Mn^{2+}, Ni^{2+}, Pb^{2+}, Co^{2+}, Zn^{2+}, Fe^{3+}, $Cs^{2+}$and Mg^{2+}, of each at a concentration of 15.0 μg mL^{-1} were reacted with a PB-CQDs solution. Fig. 7 A displays the relative change in PL intensity of PB-CQDs in the presence of various metal ions. Fe^{2+}, Mn^{2+} and Hg^{2+} ions cause the slight PL changes (defined as the relative change of PL intensity in 60–80% as compared to blank) which can be attributed to the nonspecific interactions between the carboxylic and/or amine groups and the metal ions. Among these metal ions, Cu^{2+} displays the strongest PL quenching effect on PB-CQDs, assigning to the special coordination between the Cu^{2+} ions and the phenolic hydroxyl and/or amine groups of PB-CQDs which has been widely used for the detection of Cu^{2+} ions or colored reactions in traditional organic chemistry. Such a specific fluorescence quenching effect may originate from the strong interactions between Cu^{2+} ions and the surface groups of PB-CQDs which transfer the photoelectrons from PB-CQDs to Cu^{2+} ions. These results clearly demonstrate that the PB-CQDs-based Cu^{2+} sensor is highly selective to Cu^{2+} over the other metal ions. Fig.8. Fluorescence responses of CCDs to the addition of different metal ions under acetate buffer (pH 4.3). The concentration of each metal ion is 15 μg $mL^{-1.}$ F_0 and F correspond to the fluorescence intensities of PB-CQDs at emission wavelength of 440 nm excited at 320 nm in the absence and presence of metal ions, respectively.

As shown in Fig. 3, the fluorescence intensity of PB-CQDs gradually decreased with the Cu^{2+} concentration. The fluorescence intensity displayed liner responses with the Cu^{2+} concentrations, with two linear ranges of 0.0□14.0 μg mL^{-1} and 15.0□25.0 μg mL^{-1} (Fig.3B). The linear equations are $F/F0$ =0.9827□0.8485C_{Cu2+} (R = 0.993) and $F/F0$ = 0.7565–0.0033 C_{u2+} (R = 0.996), respectively.

REFERENCES

[1] W. B. Lu, X. Y. Qin, S. Liu, G. H. Chang, Y. W. Zhang, Y. L. Luo, A. M. Asiri, A. O. Al-Youbi, X. P. Sun, Economical, green synthesis of fluorescent carbon nanoparticles and their use as probes for sensitive and selective detection of mercury(II) ions, Anal. Chem. **2012**, 84, 5351-5357.

[2] J. Jiang, Y. He, S. Li, H. Cui, Amino acids as the source for producing carbon nanodots: microwave assisted one-step synthesis, intrinsic photoluminescence property and intense chemiluminescence enhancement. Chem Commun **2012**, 48, 9634–6.

[3] Y. Wang, A. Hu, Carbon quantum dots: synthesis, properties and applications, J. Mater. Chem. C **2014,** 2, 6921–6939.

[4] C. Liu, P. Zhang, X. Zhai, F. Tian, W. Li, J. Yang, et al., Nano-carrier for gene delivery and bioimaging based on carbon dots with PEI-passivation enhanced fluorescence, Biomater. **2012,** 33, 3604-13.

Figures:

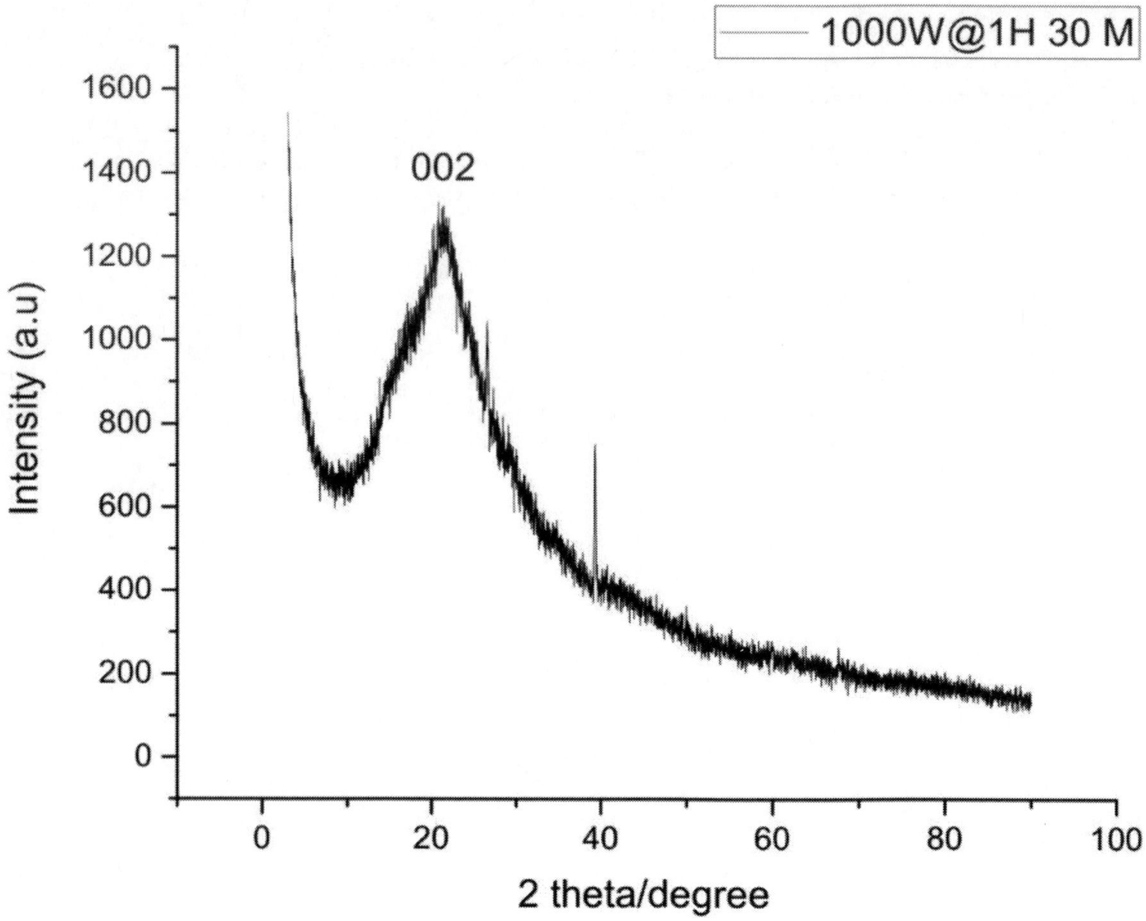
1000W@1H 30 M
002
Intensity (a.u)
1600
1400
1200
1000
800
600
400
200
0
0
20
40
60
80
100
2 theta/degree

Fig 1. FTIR spectra of as-prepared C-dots from Pine Bark

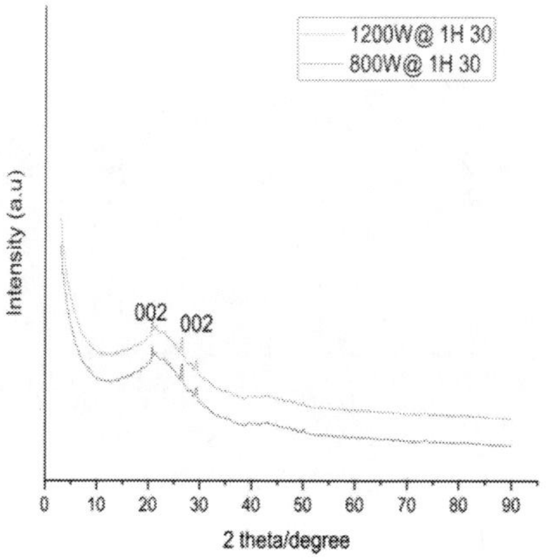

(a)

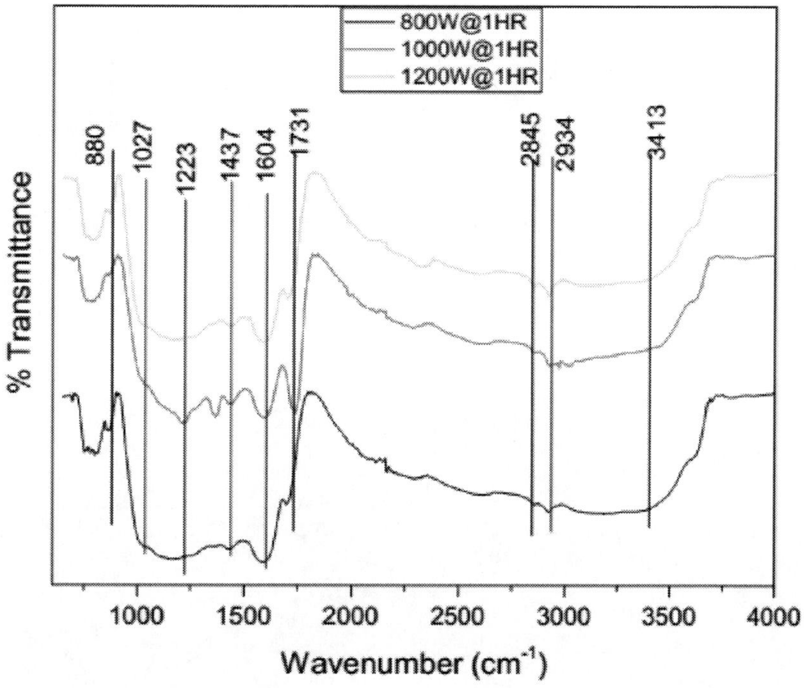

(b)

Fig. 2a & b. XRD pattern of C-dots.

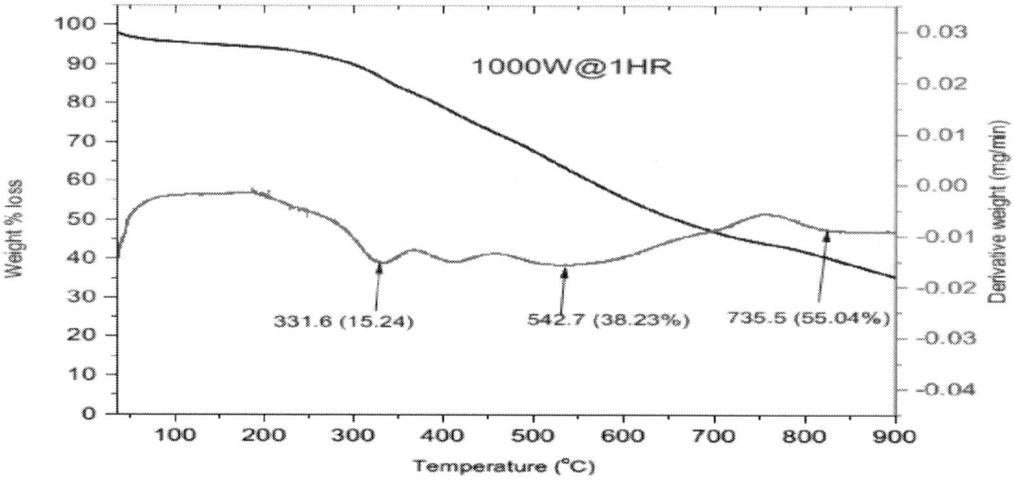

Fig 3. Thermal curves for C-dots under N_2 atmosphere.

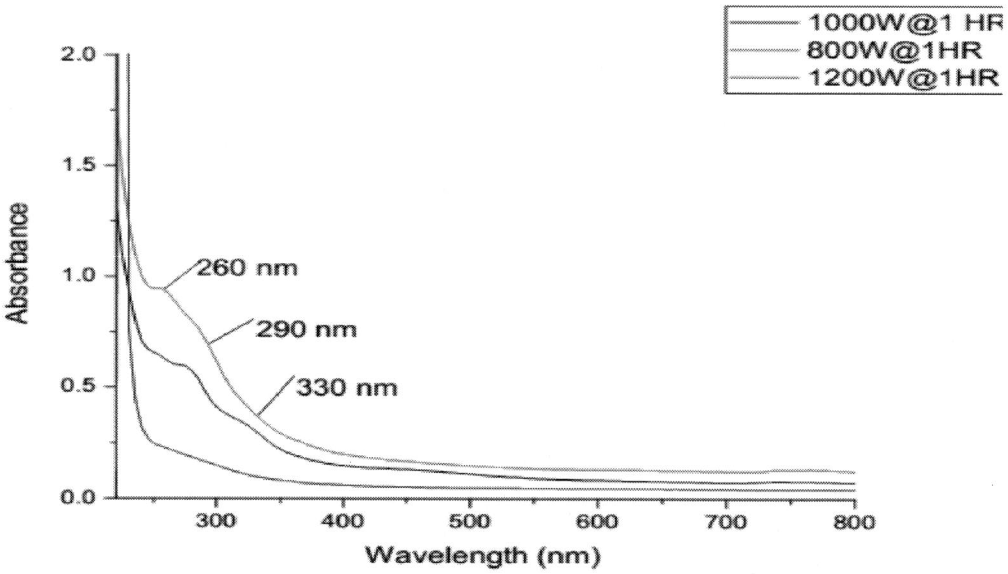

Fig 4. UV-vis absorption of C-dots.

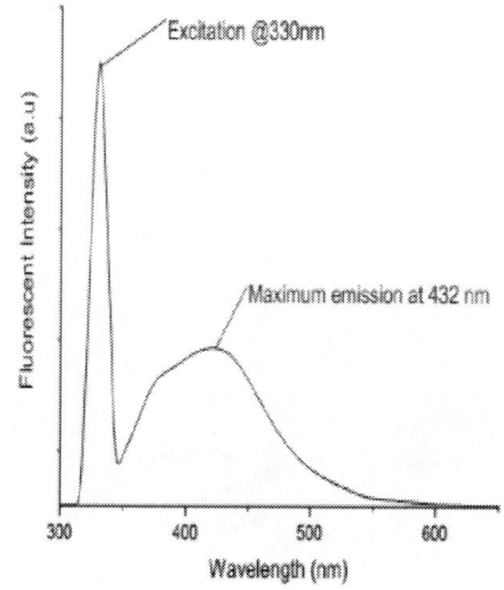

(a)

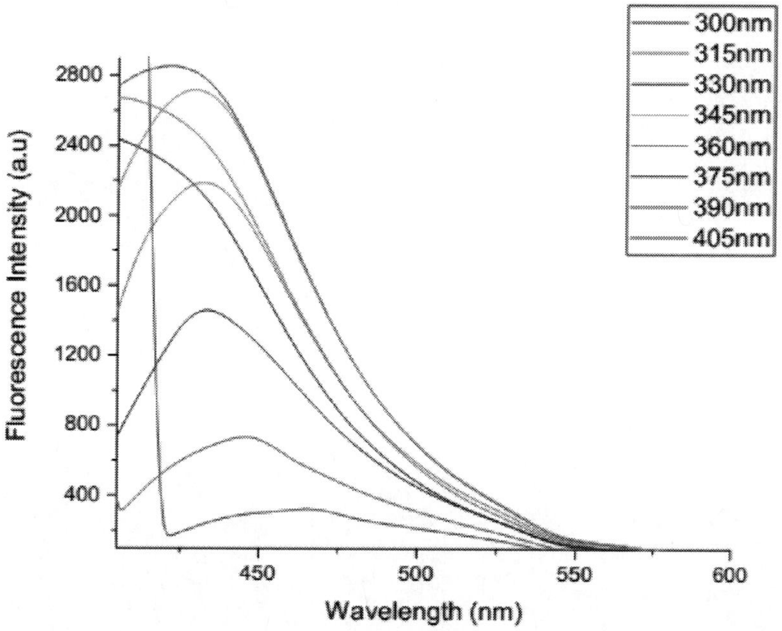

(b)

Fig 5a. PL excitation and emission wavelength (b) PL emission spectra at different excitation wavelength from 300 to 405 nm in 15 nm increments.

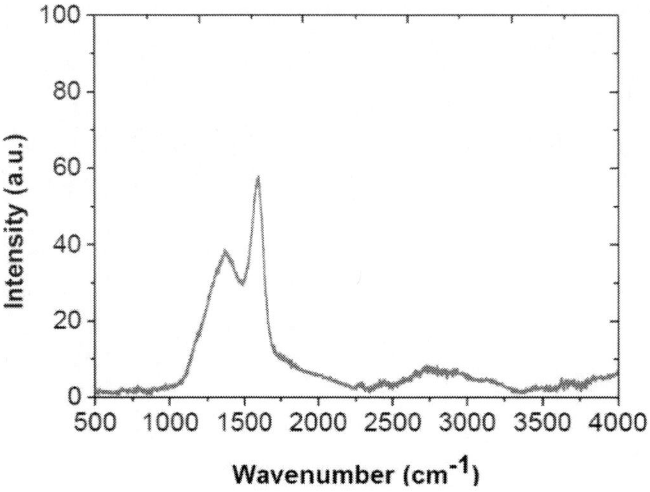

Fig. 6. Raman Spectra of PB-CQDs

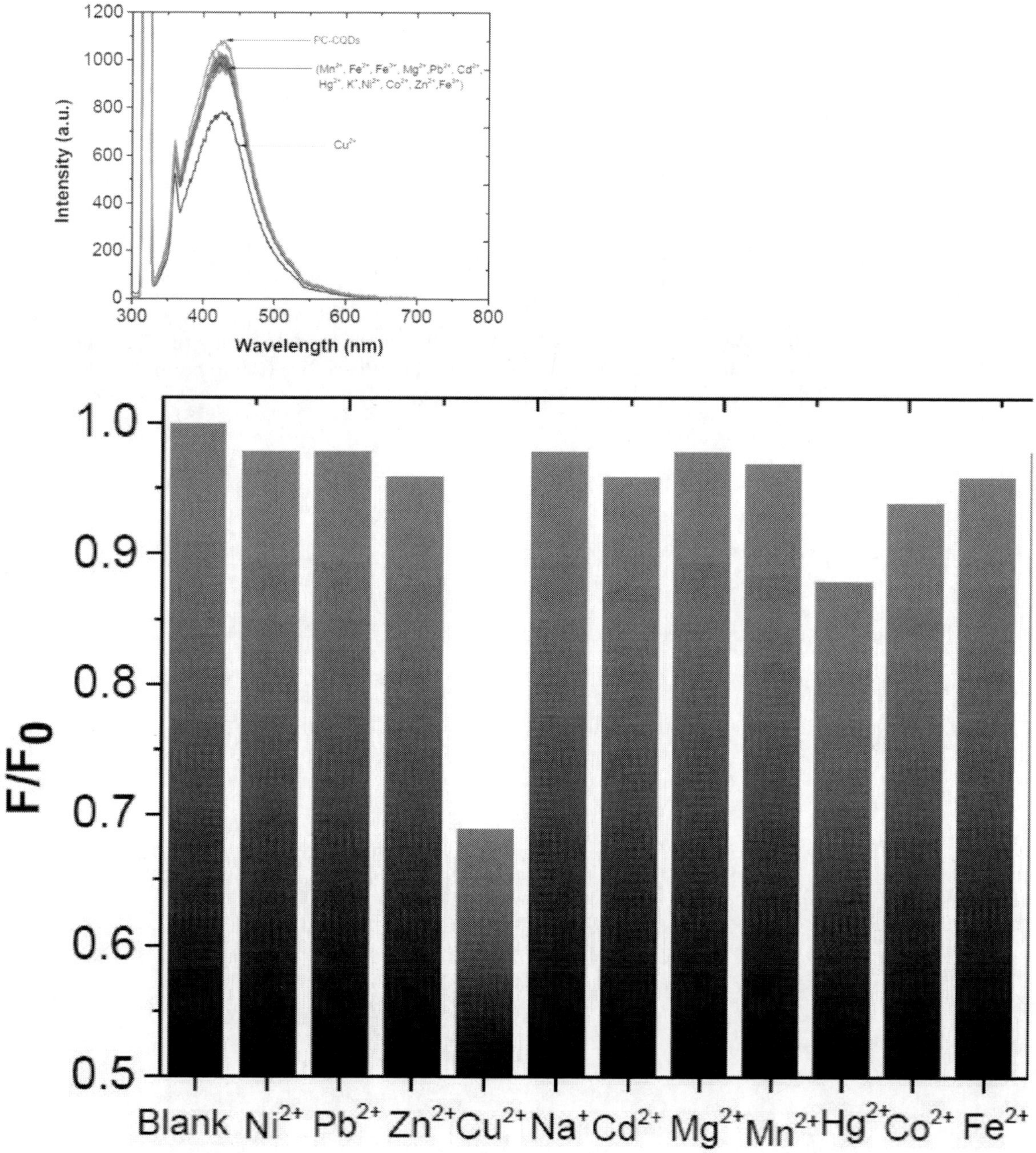

Fig.7. Fluorescence responses of CCDs to the addition of different metal ions under acetate buffer (pH 4.3).

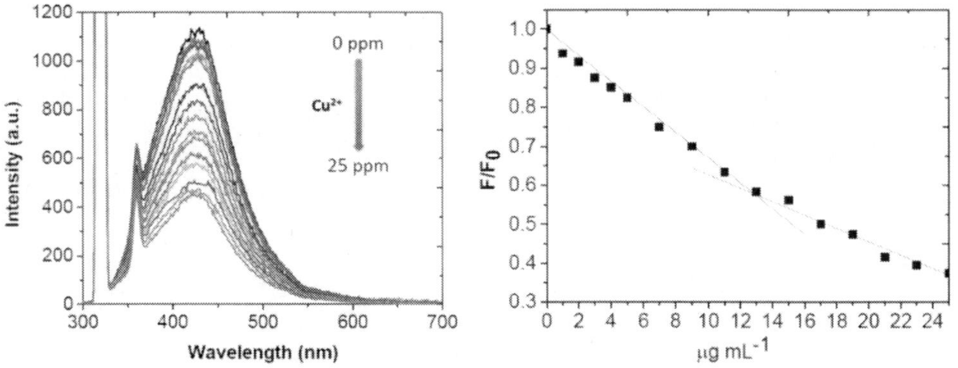

Figure 8. Fluorescence titrations of the y-CDs with Cu^{2+} from 0 to 25 μg mL^{-1}. (Influence of Cu^{2+} concentration on the fluorescence intensity of PB-CQDs (top to bottom: 0, 1, 2, 3, 4, 5, 7, 9, 11, 13, 15, 17, 19, 21, 23 and 25.0 μg mL^{-1})

443095 Experimental and Numerical Study of Transport Phenomenon of Falling Film Flow with Different Heating Methods

Abstract:

Falling film flow is a common phenomenon in chemical industry. It is widely used in chemical instruments, such as structured packed columns and falling film evaporators. An experiment was implemented to study the mass transfer phenomenon of binary-mixture (ethanol-water) falling film flows with different heating methods (uniform or concentrated heating). In order to study the detailed flow and mass transfer behaviors, a 2-D multiphase computational fluid dynamics (CFD) model based on volume of fluid (VOF) was developed. This model divides mass transfer into two parts: one is caused by the vapor-liquid equilibrium (VLE) and the other is caused by the heat flux. Results from simulation are inconsistent with the experimental data. Both experiments and simulations show that the ethanol concentration of the vapor with concentrated heating is 1.5% higher than the one with uniform heating.

443774 Thermodynamic Anaysis of Dehydration of n-Propanol Using Heterogeneous Azeotropic Distillation

Abstract:

Heterogeneous azeotropic distillation is a widely popular process in chemical process industries for the separation of binary azeotropic mixtures into their constituent pure components. It involves the addition of an entrainer which is partially miscible and forms a heterogeneous azeotrope preferably with one of the original components. This is a powerful technique which combines several physical phenomena such as presence of azeotropes and liquid-liquid immiscibilities, and is used widely in industries to enhance difficult or intractable distillation separations. For the design of heterogeneous azeotropic distillation columns, a detailed study of the heterogeneous region is crucial to plan the distillation sequence, which requires experimental vapor-liquid-liquid equilibrium (VLLE) data.

The present study focuses on the investigation of the heterogeneous region of the well known industrially available azeotropic mixture n-Propanol-Water system using Propyl acetate as entrainer. The VLE data of the system was generated by means of modified Othmer still and the still is modified using a probe type ultrasonic homogenizer to account for the heterogeneity. Isobaric vapor- liquid- equilibria and vapor - liquid - liquid - equilibria data at atmospheric pressure have been obtained for the ternary system n-Propanol-water-propyl acetate and the constituent binary systems. The equilibrium data has been correlated using NRTL, UNIQUAC and UNIFAC models and compared with the experimental data.

444269 Functionalization of OH-Terminated Nanoparticles By Cyclic Azasilanes

Abstract:

Colloidal suspensions of functionalized nanoparticles have been suggested for enhanced oil recovery, formation damage remediation, flow assurance, and water treatment operations. Surface modified nanoparticles can also be used as tracers in reservoir studies, active media in thermal management fluids, and fillers in high performance elastomers. Conventional functionalization of nanoparticles with alkoxysilanes and chlorosilanes is limited by needs for catalytic amount of water for the reaction to proceed, requirement of multiple hours to fully saturate the surface, and release of by-products upon reaction completion. Monolayer deposition with a high density of functional groups represents a particularly difficult challenge.

Our ongoing studies indicate that application of cyclic azasilanes that have a strained ring structure with adjacent Si and N centers provides highly efficient functionalization of hydroxylated silica, alumina, magnetite, and diamond nanoparticles which is catalyst-free and by-product-free. The data on synthesis and characterization of functionalized nanoparticles as well as possible reaction mechanism of functionalization will be discussed.

443776 Study of Ionic Liquids As Entrainers for the Extractive Distillation of *N*-Propanol and *N*-Propyl Acetate Binary Azeotropic Mixture and the Isobaric VLE Studies of the System

Abstract:

Distillation is a widely accepted technique for the separation of binary liquid mixtures. Separation by simple distillation is not possible for binary mixtures that form azeotropes. One of the commonly used methods to separate azeotropes is extractive distillation. It is the process of adding an entrainer that can alter the relative volatility of the mixture and thereby breaking the azeotropic point. Ionic liquids on account of their physiochemical properties as well as the recyclability have been used as a preferred replacement for the conventional entrainers. Extractive distillation with ionic liquids has the advantages of both liquid solvent (easy operation) and solid salt(high separation ability). The corrosion and dissolution problems by using inorganic salts can be rectified by these non-volatile, chemically stable, high electrical conductive molten salts.

The azeotropic system under consideration in the present study is *n*-propanol - *n*-propyl acetate mixture. These compounds which constitute the products/reactants of esterification reaction of n-propanol form a minimum boiling azeotrope at 94.7°C. *n*-propyl acetate has been included in a list of volatile acetates as an ecologically benign CFC-free solvent for aerosol sprays.

Vapour liquid equilibrium(VLE) measurements of *n*-propanol and *n*-propyl acetate are performed at atmospheric pressure in a modified Othmer still which is a continuous distillation still in which vapour phase is recirculated. The most suitable ionic liquids for shifting of azeotropic composition of the system are selected using COSMO-RS model with suitable combinations of anions and cations. The isobaric VLE of the system with [EMIM][HSO_4] is obtained and compared with the existing data. The influence of alkyl chain length on the performance of ionic liquids is also analysed with regard to the previous studies. The effect of polarity is the most important factor on the behaviour of ionic liquid solvation in alcohols. Experimental VLE data were correlated using modified UNIFAC model.

444280 Copper-Phyllosilicate Core-Shell Nanocatalysts with Balanced Active Sites for Carbon-Oxygen Hydrogenolysis Reactions

Abstract:

Copper-Phyllosilicate Core-shell Nanocatalysts with Balanced Active Sites for Carbon-Oxygen Hydrogenolysis Reactions

Hairong Yue, Weimao Fang, Changmin Liu, Jiankang Ying

College of Chemical Engineering, Sichuan University, Chengdu, China

Email: hryue@scu.edu.cn

Abstract

Hydrogenolysis of carbon-oxygen (C¯CO) bonds (e.g., esters, ethers, furfural, and CO_2) has emerged as a versatile synthetic tool in organic methodology, as it could produce a variety of products (e.g., chemicals, fuels, and polymers). Copper-based catalysts have been intensively explored for hydrogenation reactions as the copper sites account for the selective hydrogenation of carbon-oxygen bonds and relatively inactive for the hydrogenolysis of carbon-carbon bonds. Work on understanding the active sites of copper catalysts for hydrogenation reactions indicted that both Cu^0 and Cu^+ species were crucial to the activity of Cu-based catalysts. However, researchers were not able to establish a relationship between activity and Cu^0/Cu^+ active species since the conventional Cu-based catalysts is the deactivation by metal particle growth and unstable surface Cu^0 and Cu^+ active species in the strong reductive H_2 and oxidizing carbon-oxygen atmosphere.

We renctly present on efficient approaches for fabrication of a series of copper-phyllosilicate core-shell nanocatalysts and nanoreactors with balanced and stable Cu^0 and Cu^+ active species. We chose the hydrogenation of dimethyl oxalate, ethylene cyclicarbonate and as probe reactions for a better understanding of properties of active sites and the structure-activity relationship. The results indicated a synergy between copper and the oxide components and the balanced surface Cu^0 and Cu^+ species can greatly improve the catalytic C-O hydrogenolysis performance of the catalysts. These nanocatalysts with balanced and stable Cu^0 and Cu^+ active species, confinement effect, intrinsic high surface of Cu^0 and Cu^+ and unique tunable tubular morphology, has potential applications in the high-temperature hydrogenation reactions.

Keywords: core-shell nanocatalysts, Hydrogenation reaction; copper catalyst

References

1. Yue, H.; Zhao, Y.; Ma, X.; Gong, J.: Ethylene glycol: properties, synthesis, and applications. *Chem Soc Rev* 2012, 41, 4218-4244.

2. Gong, J.; Yue, H.; Zhao, Y.; Zhao, S.; Zhao, L.; Lv, J.; Wang, S.; Ma, X.: Synthesis of Ethanol via Syngas on Cu/SiO2 Catalysts with Balanced Cu^0-Cu^+ Sites. *J Am Chem Soc* 2012, 134, 13922-13925.

3. Zhao, S.; Yue, H.; Zhao, Y.; Wang, B.; Geng, Y.; Lv, J.; Wang, S.; Gong, J.; Ma, X.: Chemoselective synthesis of ethanol via hydrogenation of dimethyl oxalate on Cu/SiO_2: Enhanced stability with boron dopant. *J Catal* 2013, 297, 142-150.

4. Yue, H.; Ma, X.; Gong, J.: An Alternative Synthetic Approach for Efficient Catalytic Conversion of Syngas to Ethanol. *Acc Chem Res* 2014, 47(5):1483-1492

5. Yue, H.; Zhao, Y.; Zhao, L.; Lv, J.; Wang, S.; Gong, J.; Ma, X.: Hydrogenation of dimethyl oxalate to ethylene glycol on a Cu/SiO2/cordierite monolithic catalyst: Enhanced internal mass transfer and stability. *AIChE J* 2012, 58, 2798-2809.

6. Yue, H.; Zhao, Y.; Zhao, S.; Wang, B.; Ma, X.; Gong, J.: A copper-phyllosilicate core-sheath nanoreactor for carbon¯Coxygen hydrogenolysis reactions. *Nat Commun* 2013, 4, 2339.

443819 Improving High Shear Rate Friction Reduction Behavior of Polyacrylamide-Based Fracturing Fluids Using Xanthan Gum

Abstract:

The main component of slickwater in fracturing low permeability formations is a high molar mass polymer such as polyacrylamide (PAM), which acts as a drag reduction agent in turbulent flow. Polyacrylamide does not degrade in the environment and also suffers from mechanical degradation at high turbulent flow rates. Five different concentrations of PAM as well as mixtures of polyacrylamide/xanthan were prepared to investigate the possibility of improving shear stability of PAM. Drag reduction measurements were performed in a flow loop and friction factors at different flow rates were calculated using rheological parameters. Based on the observations, Xanthan (XG) is not as good a friction reducer as PAM, even at three-fold higher concentration. XG slightly improved the shear stability in mixed polymer solutions at low concentrations. But at high concentrations, XG significantly improves the friction reduction behavior of PAM at nonionic and ionic environments.

Stimuli-responsive Rubber in Oil and Gas Applications

Rostyslav Dolog, Sankaran Murugesan, and Valery Khabashesku

Baker Hughes, 14990 Yorktown Plaza Dr, Houston, Texas, 77040, USA

ABSTRACT

A problem that exists today in the oil & gas industry is the loss of mechanical properties by reactive rubber seals due to swelling, and the limited space in downhole well applications to accommodate larger swellable packing elements.

This paper presents cost-effective, reinforcing, stimuli-responsive fillers, such as certain classes of fly ash, for use in rubber composite materials in various oil and gas applications. Apart from cost effectiveness, these fillers can also enable stimuli-responsive effects, as well as improved mechanical properties (e.g. modulus) for rubber composites when swelled in water. Fly ash (Class C) is a common component of Portland cement which was used in the literature to achieve a stimuli-responsive effect when added to Hydrogenated Nitrile Butadiene Rubber (HNBR) as filler. Exposure to water brines can cause a filler to promote an increase in pH which enables the hydrolysis of HNBR nitrile groups to carboxyl groups, thus creating an ionomer. This process is followed by neutralization of the new ionomer with metal ions, and formation of ionic crosslinks, further improving mechanical properties of the rubber.

This paper demonstrates that incorporation of fly ash as a reinforcing filler allows development of swellable seals that can retain, or even increase, their modulus and stiffness upon swelling. These characteristics allow avoiding the loss of mechanical properties by seals due to swelling and also eliminating the need for large swellable packing elements that occupy additional space downhole.

KEYWORDS: Elastomers, stimuli-responsive, fly ash, composites

INTRODUCTION

There is a growing demand for high performance elastomers in the oil & gas industry. Rubber compounds are used for various sealing applications, packer elements, stators for pumps, etc. In general, rubber formulations are comprised of elastomer matrix, curing system, antioxidants, plasticizers, along with reinforcing filler, which provides the compounds with their strength. The reinforcing filler is usually represented by carbon black and, to a lesser extent, silica. Along with other factors, filler mechanical properties and concentration generally define hardness, modulus and tensile strength of the elastomer compound. Throughout the years, many different materials have been tried as alternative fillers for rubber composites, including nanomaterials (CNTs, graphene, nanofibers), recycled rubber particles, and ash/fly ash. In the case of ordinary carbon black, the filler serves not only for reinforcement, but also as a cost reducer, because raw elastomer is often a more expensive component. Mechanical properties of the composites are affected by the aspect ratio of the filler, quality of the dispersion, and interface between blend components. The latter determines the effectiveness of load transfer from polymer matrix to the reinforcing filler. High performance compounds can be designed by incorporating well dispersed fine particles with a high aspect ratio and good compatibility with the polymer matrix.

In this context, the fly ash is a cheaper, low reinforcing filler alternative to carbon black, which will help reduce cost of the compounds for "non-demanding" sealing applications requiring large amounts of raw materials. Fly ash is produced in large quantities as a byproduct of coal burned at power plants, and can be considered as a recycled material. Fly ash costs several cents per pound, while carbon black, depending on the grade, is usually one order of magnitude more expensive. Hemmings et al[1] demonstrated that fly ash can be used as filler for polymeric composites with enhanced mechanical properties. Adoption of this technology in the oil and gas industry could change the design approach for reactive rubber compounds. Further functionalization of fly ash through a cost-efficient process could ensure compatibility with the rubber matrix, while promoting filler dispersion and ensuring an efficient load transfer. Such functionalization's include, but are not limited to, surface modification with fatty acids, dicarboxylic acids or phosphoric acid. The reasoning for surface modification is to attach functional groups with aliphatic chains to fly ash particles to make them compatible with aliphatic rubber chains, or to attach acidic groups on their surface that can interact with nitrile groups of HNBR. Initial attempts towards chemical functionalization of fly ash have been made by Seal et al for treatment of offshore oil spills.[2]

Apart from cost effectiveness, composites with fly ash can show stimuli-responsive properties. It has been shown that adding Portland cement to the HNBR matrix improves its mechanical properties in the swollen state by water uptake.[3] Exposure to water causes Portland cement to grow in volume and mass, and promotes an increase in pH which enables the hydrolysis of HNBR nitrile groups to carboxyl groups, thus creating an ionomer. This process is followed by neutralization of the new ionomer with metal ions and formation of ionic crosslinks, further improving mechanical properties of the rubber. Fly ash (Class C) is a common component of Portland cement, which contains a large variety of metal oxides, and is capable of increasing the

[1] Filler Comprising Fly Ash For Use In Polymer Composites. Hemmings et.al., US 6,916,863 B2, July 2005.

[2] Hydrophobic High Surface Area Zeolites Derived from Fly Ash for Oil Spill Remediation, Seal et.al., Environ. Sci. Technol., 2013, 47 (11), pp 5843–5850

[3] Stimuli-Responsive Cement-Reinforced Rubber, Musso et. al., ACS Appl Mater Interfaces. 2014 May 14;6(9):6962-8.

pH of the water solution by hydrolysis, which has a potential to achieve a stimuli-responsive effect when added to HNBR as filler.

Degradation of tensile properties is a common problem for swellable packers and other sealing elements. By selectively using certain classes of fly ash as effective reinforcing filler, one can develop swellable seals that can retain or even increase their modulus and stiffness upon swelling. Therefore, there is a potential to minimize or even avoid the loss of mechanical properties by water swellable reactive packing elements due to swelling. This provides an opportunity to decrease the size of these elements, which are often represented by rather long sections, and hence occupy a significant amount of space downhole. In this paper we have studied the reinforcing effects of fly ash fillers with HNBR compounds as well as stimuli responsive properties of these compounds.

EXPERIMENTAL SECTION

Rubber compounds T1, T2, T3, and T4 are elastomers based on a custom-built peroxide curing system. Formulation F2 was prepared by adding 15 wt. % of fly ash class C to compound F1, which is a commercial swellable NBR sulfur-cured compound.

T series compounds were prepared using a C.W. Brabender® internal mixer operated at 60 RPM for 20 minutes at ~160°F and then molded into sheets and cured at 340°F for 30 minutes in a hot press. Compound F2 was prepared by mixing fly ash with compound F1 in the Brabender mixer for 6 minutes at 60 RPM at ~100°F. Compound F1 is a commercial water swellable formulation that was used as received. F series compounds were cured at 300°F for 60 minutes in a hot press.

Tensile properties of rubber samples were tested using dog bone specimens, according to the ISO37 standard, at room temperature and at 350°F using an MTS® testing system.

Swelling tests in 3% KCl brine and LVT 200 oil were performed to assess the effect of fillers on the chemical resistance of elastomers. Standard one-inch rubber buttons were tested at 200°F at ambient pressure for 6 days. Thermal properties were studied using TGA Q50 and DSC Q2000 instruments.

To analyze the stimuli-responsive effect introduced by fly ash fillers, rubber buttons were aged for 10 days in 3% KCl in the oven at 200°F and then removed from the solution, dried for 30 minutes in the oven and then for 3 hours at room temperature. Samples were later tested in the MTS testing system in compression mode.

Viscoelastic properties of stimuli-responsive samples, based on water swellable compounds, were accessed using the RSA G2 dynamic mechanical analysis (DMA) instrument. This experiment was conducted in a temperature ramp mode of 2°C/minute at 1Hz frequency and 0.2% amplitude strain. A Nicolet 6700 Fourier Transform Infrared (FTIR) spectrometer was used to confirm neutralization of carboxylic groups by metal cations.

RESULTS AND DISCUSSION

The reinforcing properties of fly ash were evaluated at room temperature (Table 1) and at 350°F (Table 2) and compared with N550 carbon black. Sample T3, reinforced with fly ash, demonstrated inferior tensile properties and hardness compared to a compound filled with carbon black. However, at room temperature, fly ash was capable of delivering a reinforcing effect to the rubber compound without sacrificing its elasticity, e.g. the ultimate elongation of elastomer compound remained virtually unchanged after incorporation of 50 parts per hundred rubber (PHR) of fly ash, while its modulus and tensile strength have increased. Therefore, potentially, one can substitute carbon black with high loadings of fly ash in certain applications, thus further decreasing the cost of the rubber product. The summary of mechanical properties is provided in Table 1, E_x denotes modulus of a compound at certain strain (x), i.e. stress at strain.

Sample	N550 CB (PHR)	Fly ash Class F (PHR)	Tensile strength (psi)	E_{25} (psi)	E_{50} (psi)	E_{100} (psi)	Elongatio n (%)	Hardness Shore A
T1	0	0	829	71	97	121	472	48
T2	50	0	3402	165	273	744	290	68
T3	0	50	1176	94	127	166	467	54

Table 1. Room temperature tensile properties and hardness of elastomer samples

The relatively low increase in modulus and hardness of T3 compared to T1 at room temperature can be explained by a low aspect ratio of fly ash and weak interactions between blend components. The latter problem can be solved by functionalization of fly ash particles.

The general data trends from the high temperature tensile tests closely resemble the results of the tensile test conducted at room temperature. However, at high temperature, rubber specimens break at much lower elongation, which leads to a great decrease in the measured tensile strength. This behavior is typical for elastomers and limits their application to a certain maximum service temperature and is also due to the fact that high temperature tests are conducted at conditions that push the limits of the material. The modulus values at 15% (E_{15}) and 20% (E_{20}) strain are not provided herein, because half of the tested specimens were broken before they reached such strains.

The superiority of carbon black as an efficient filler became even more pronounced at high temperature. However, incorporation of fly ash doubled the tensile strength and increased the elongation at break by one-third when compared to the unfilled elastomer.

Sample	E_5 (psi)	E_{10} (psi)	E_{15} (psi)	E_{20} (psi)	Tensile strength (psi)	Elongation (%)
T1	31	46			91	23
T2	61	130	194	261	832	39
T3	42	73	98	119	173	30

Table 2. Summary of high temperature (350°F) tensile data for elastomer samples

Thermal properties of samples T1-T3 were assessed with Differential Scanning Calorimetry (DSC) and Thermogravimetric analysis (TGA). TGA allows evaluating thermal stability of samples and to verify the actual composition of a sample from the ash content. The results of the DSC and TGA analysis are summarized in Table 3. During TGA, samples were heated to 600°C at 10°C/minute rate in a nitrogen (N_2) environment, and subsequently heated at 10°C/min to 600 °C in air to burn off carbon-based fillers. This allows decomposing of organic matter first by applying heat, and then burn the remaining carbon-based material in air atmosphere.

The initial degradation of the polymer started at 200°C and progressed to complete decomposition of aliphatic chains in the temperature range of 420-500°C with a highest rate of decomposition at ~460°C. Sample T2 incurred the second rapid mass loss after 600°C due to burning of the carbon black, which comprised most of the ash content at 600°C in a N_2 environment. The DSC experiment demonstrated the glass transition temperature of the elastomer remained largely independent of the type of filler - the respective DSC heating scan was conducted with a 10°C/minute temperature ramp after precooling the sample to -50°C.

Sample	W_{600} (%)	W_{800} (%)	$T_{deg}1$, °C	$T_{deg}2$, °C	T_g, °C
T1	2.4	2.4	452.8	-	-29
T2	29.8	2.7	458.7	616.5	-29
T3	32.8	32.7	447.3	-	-28

Table 3. TGA/ DSC results for elastomer samples

Table 3 contains ash contents (W) at 600°C and 800°C, degradation temperatures (T_{deg}) which represent the temperature of the highest rate of decomposition during TGA experiment, and glass transition temperature (T_g) of the samples, as determined by DSC scan.

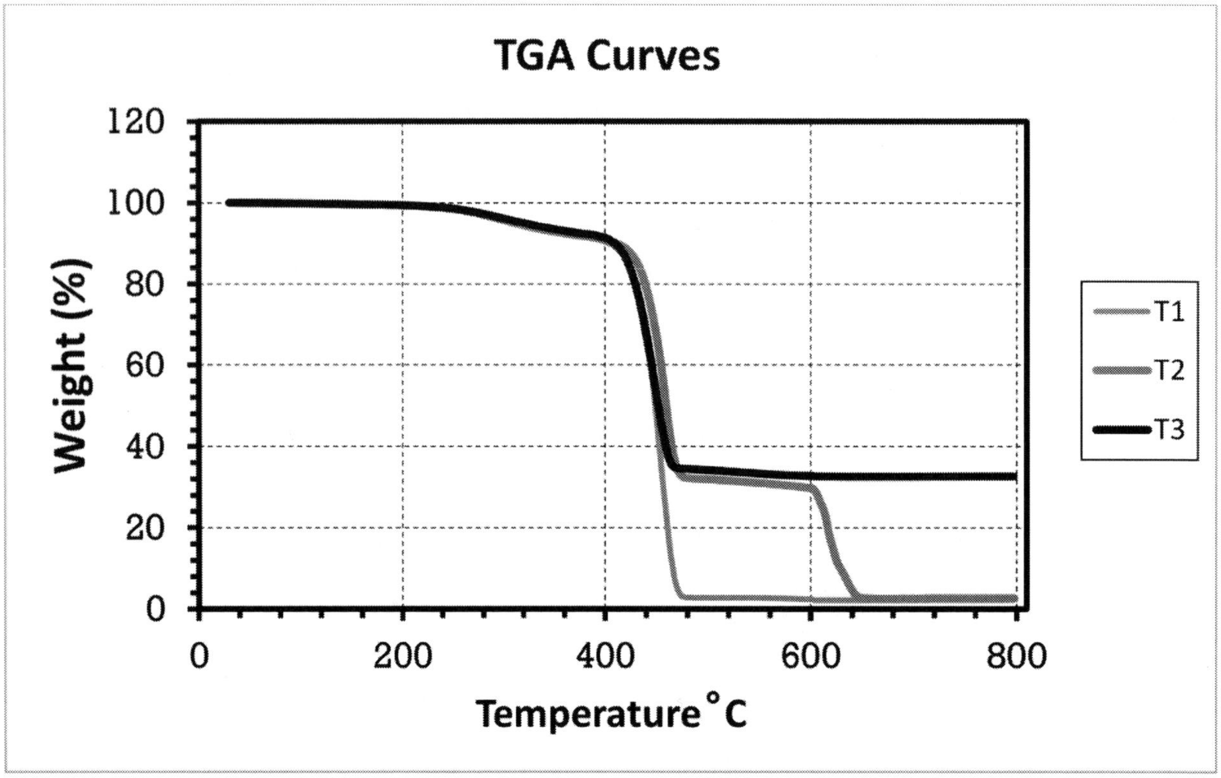

Figure 1. TGA Curves for samples T1, T2, and T3

Sample \ Testing fluid	3% KCl brine	LVT 200	Specific gravity
T1	0.6	13.0	0.96
T2	0.3	11.6	1.13
T3	0.8	9.1	1.18

Table 4. Volume increase (%) of elastomer samples due to swelling for 6 days at 200°F

Musso et. al. developed a stimuli-responsive HNBR composite reinforced with Portland cement that was increasing its stiffness and modulus when swelled in water[3]. Fly ash (Class C) is a common component of Portland cement, which contains the components necessary to achieve a similar effect. Therefore, one can avoid the loss of mechanical properties by water swellable packing element seals upon activation. To test this hypothesis, we studied the effect of aging in water brines on the properties of elastomers with fly ash fillers.

Rubber samples were molded and cured into standard size rubber buttons (cylinders of ~0.25" height and ~1" diameter). These buttons were aged for 10 days in 3% KCl brine in the oven at 200°F, removed from the solution, dried for 30 minutes in the oven and for 3 hours at room temperature. The swelling of both compounds was close to ~1% by volume (Table 4). Samples were later tested in an MTS machine in compression mode. Each button was compressed to 70% strain, buttons remained intact during the experiment, and visually all appeared to have a certain level of compression set at the end of the experiment. The results of the experiments are summarized in Table 5 and Figure 2. Only one button was tested for each sample. The provided compression modulus values (C_n) represent values of stress at a percentage strain (n), e.g. C_5 is stress at 5% compression strain.

Compound T4 was reinforced by fly ash Class C, which contains larger weight fraction of CaO than fly ash Class F (~25% vs ~10-15% respectively) and is expected to produce a higher pH and closely resemble cement that was used to prepare stimuli-responsive rubber in the original paper by Musso et.al.[3] The results demonstrate that exposure to water (3% KCl) at high temperatures can improve the modulus for the compounds filled with fly ash Class C (T4), while the control sample (T1) showed the opposite behavior. After aging, the compressive modulus at 50% strain (e.g. stress at 50% strain) of control sample T1 decreased by 19%, as compared to the unaged sample, while compressive strength of sample T4, reinforced by fly ash, increased by 15%. The difference in modulus values is even more drastic at 70% compression strain (-35% and 31%, respectively). This is in agreement with the hypothesis stated earlier.

Sample	Filler	Compressive modulus (psi)				
		C_5	C_{10}	C_{25}	C_{50}	C_{70}
T1	None	19	42	127	468	1810
T1 swollen	None	16	36	109	380	1185
T4	Fly Ash class C (50 PHR)	21	46	134	434	1628
T4 swollen	Fly Ash class C (50 PHR)	29	57	153	501	2125

Table 5. Summary of compressive modulus properties after 3% KCl soaking and 200°F drying

To further investigate the stimuli-responsive property of fly ash filler, the latter was incorporated into a water swellable compound, which provided for significantly more interaction of filler with water to occur. Sample F2 comprises unmodified commercial water swellable NBR formulation F1 mixed with 15 wt.% of fly ash Cass C. Sample F2 was processed into a 0.04" slab, then a rectangular sample was cut from the slab and swollen in 3% KCl solution until 100% volume increase. The F2 sample was then dried overnight in the oven, which lead to de-swelling of the sample due to solvent evaporation. Mechanical properties of the F2 sample were assessed with DMA prior to the swelling test, in a swollen state, and after drying - the data summary is shown on Figure 3.

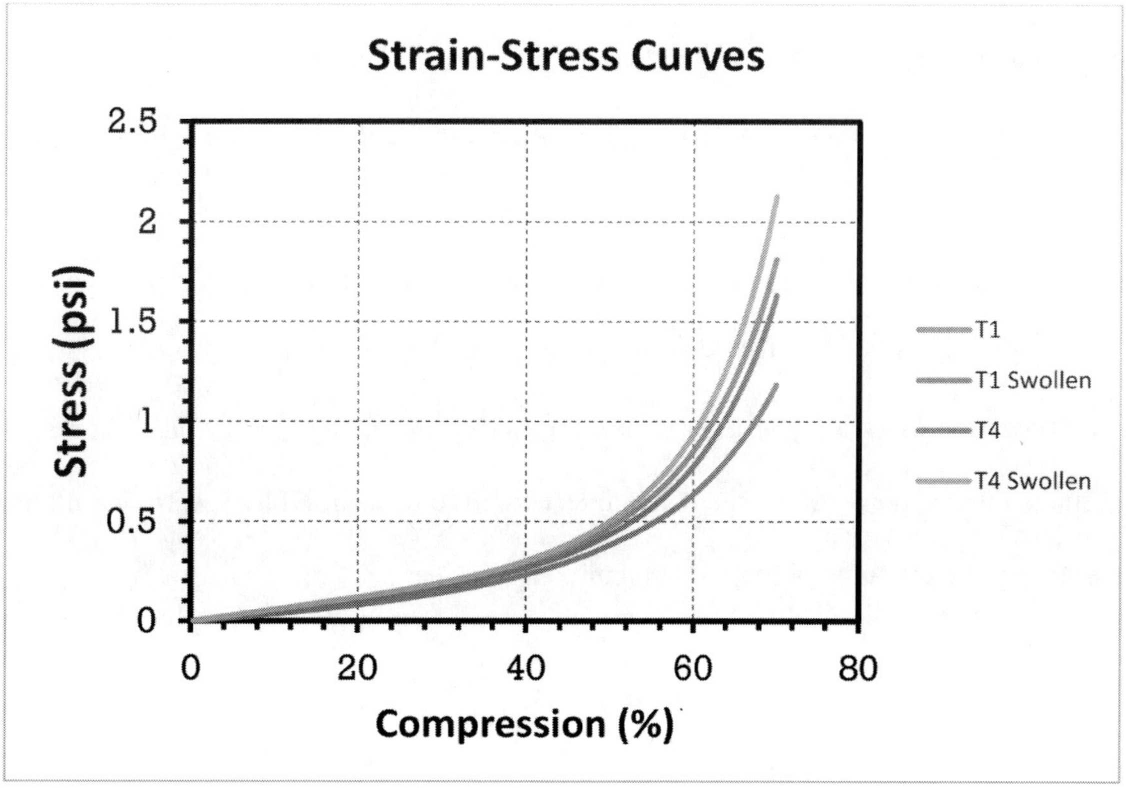

Figure 2. Stress-strain curves

The storage modulus of sample F2 after swelling and subsequent drying was about an order of magnitude higher than that of original sample F2 measured prior to swelling experiment. The storage modulus of the dried F2 sample was also less sensitive to the increasing temperature, contrary to the weakening original sample F2. The increase in the storage modulus of swollen F2 with temperature is simply due to the drying of the sample. The obtained result demonstrated that fly ash can essentially act as a stimuli-responsive material.

The significant improvement in the modulus is believed to be due to the neutralization of the swellable component of the commercial F1, and potential hydrolysis with subsequent neutralization of NBR elastomer. This is confirmed by the shift in T_g to a lower temperature range after drying as illustrated in Figure 3.

Carboxylic groups were detected in the original sample F2 prior to swelling, hence, they cannot be attributed solely to hydrolyzed HNBR, but rather to the super-absorbent polymer present in water swellable rubber compound (see Figure 4). FTIR confirms neutralization of carboxylic groups because a significant increase of the peak near ~1540 cm^{-1} and a lowering of peak intensity at ~1720 cm^{-1} have been observed. This neutralization is believed to promote the formation of ionic clusters in the compound with fly ash considered to be the source of cations. The neutralization of carboxylic groups could as well be attributed to K^+ cations present in the KCl brine solution, however, in that case, the respective peak in FTIR spectra would appear at higher wave numbers, approaching ~1600 cm^{-1}.

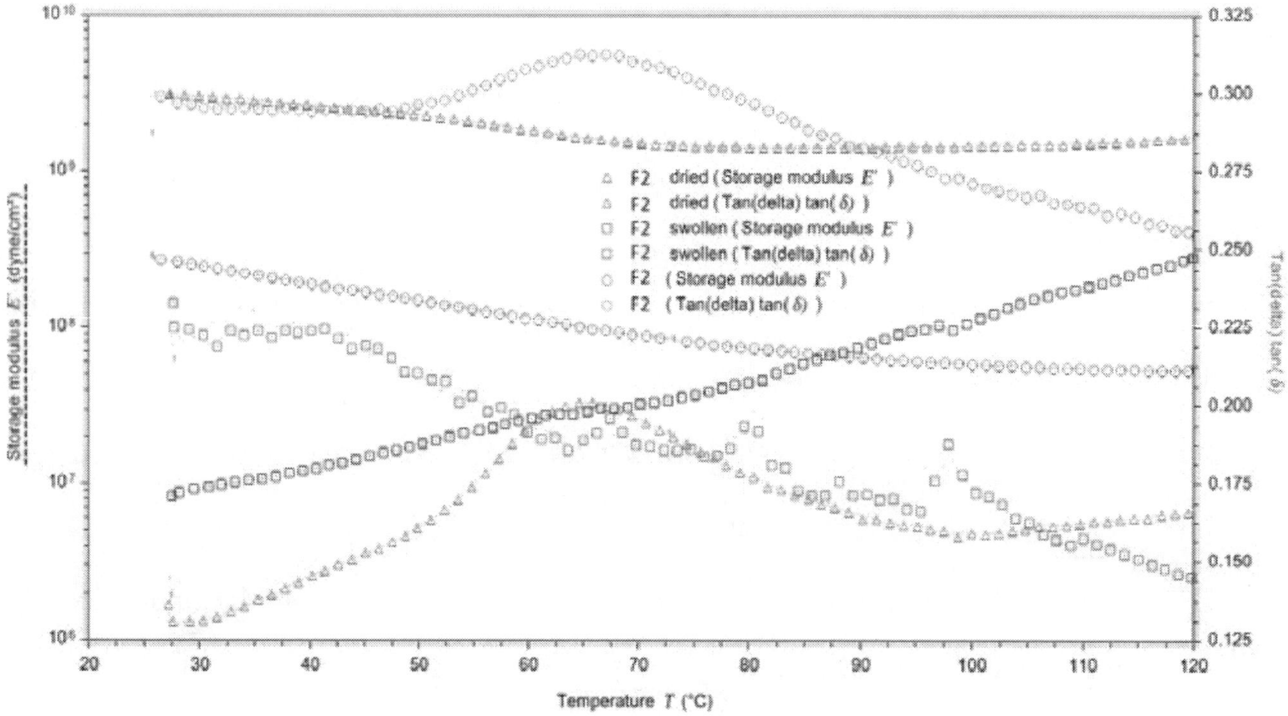

Figure 3. DMA data on F2 sample

It is important to mention that sample F2 exhibited a certain degree of brittleness after drying, which is undesired in rubbers, but this can be managed by incorporation of common ionomer plasticizers.

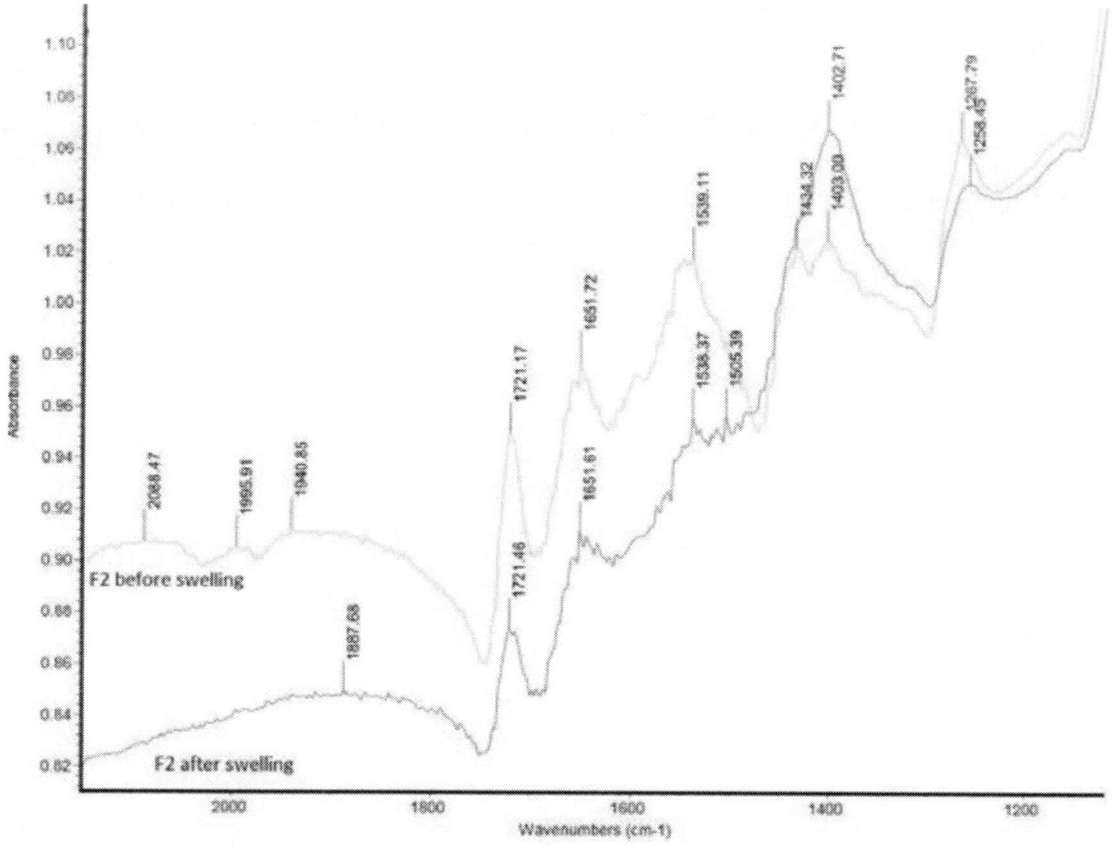

Figure 4. FTIR spectra of F2 sample taken before and after swelling

Conclusions

Fly ash demonstrated a great potential for consideration as cost effective, low reinforcing filler for downhole oil and gas applications. Although the reinforcing properties of fly ash are inferior to carbon black, incorporation of fly ash can improve the mechanical properties of HNBR elastomer compared to the unfilled raw crosslinked elastomer at both room temperature and at 325°F.

Furthermore, composites of HNBR with fly ash Class C demonstrated a stimuli-responsive effect by significantly increasing their modulus upon swelling in water brine. This effect was even more pronounced in composites of fly ash with swellable water compounds, which provided an easy access of filler to water upon swelling. A ten-fold increase in modulus was achieved in such fly ash reinforced sample, after swelling in 3% KCl water brine and drying in the oven overnight. The FTIR confirmed neutralization of carboxylic groups present in the investigated system, which contributed to overall improvement of mechanical properties of the compound upon swelling.

444093 New Development on Hoist Scheduling with the Relaxation of Cyclic Precondition

Abstract:

Multi-Stage Material Handling (MSMH) processes are designed to manufacture materials requiring a series of treatments and in industry, they are broadly used due to their capability to handle large work orders and compatibility with materials with different recipes. A typical example is electroplating process. A hoist handles the transportation of each job among units (stages) according to its processing recipe. The hoist movement plan is predetermined through optimization and this plan is repeated to perform continuity. Obviously, the ultimate goal is to maximize the productivity within each movement plan, which is equivalent to minimize the cycle time to finish one work order. The optimization process is generally called Cyclic Hoist Scheduling (CHS), since every movement plan is a cycle due to its repeatability. This precondition in term of hoist operation and unit usage is general for most cases. However, it could block the really optimal results.

In this study, this precondition has been relaxed for multi-capacity units, thus the unit usage can be more flexible. Thus, the real optimality can be found through CHS. To deal with the cycle connection issue from the relaxation of the precondition, a new module Cycle Connection Identification (CCI) is proposed. Obtained hoist schedule will be checked. Once a schedule with cycle connection problems is identified, a new cycle will be scheduled forcing the smooth connection with the first cycle. The objective of scheduling a new cycle is to match the beginning of the first cycle and the ending of the second cycle. If the two-cycle group is still with cycle connection issue, more cycles will be scheduled till the cycle group is free from cycle connection problems. Each cycle is called a sub-cycle and the constructed cycle group without cycle connection problems is called a full cycle. Note that in each sub-cycle, the work order and all transportation operations are completed.

In this paper, hoist scheduling models with/without such a precondition have been developed and proposed as mixed integer linear programming. The performance of the two models is tested through case studies and the difference is compared and demonstrated in detail.

444310 Formal Modeling of Accidents and Automatic Generation of Tutoring Scenarios for Intelligent Safety Training of Process Operators

Abstract:

With the emphasis of accident-free operations and prevention of catastrophic events, safety-training demands for the chemical plants have been on the increase globally. However, the research results on modeling of accidents and involving human operators and the advanced theory and technology for (safety) operator training have been shared limitedly among domain experts and commercial developers. Even with the recent trend of simultaneously delivering the operator training system with the construction of a new process plant, in commercial operator training systems (OTS), safety training used to implement quite limited and obvious scenarios (and the burden was almost up to the trainer), partially due to the difficulty in modeling propagations of probable accidents. By the regulatory requirements or proactive investment by companies, effective scenarios for training of safety and abnormal situation management are necessary to be improved in the aspects of the size of volume as well as the training depth. Thus, considering the emergence of enhanced virtual reality technologies and inexpensive gadgets like HMD, supporting the immersive training environment, the future efficiency of the training heavily relies on the quantity and quality of safety scenarios, which would be essential in guiding the entire training process.

In this research we first overview a government-sponsored research project developing an enhanced safety training system for chemical process operators, using haptic devices in addition to enhanced audio and visual equipment. The initial version of the system runs with a fixed set of training scenarios developed for predetermined accident scenarios. However, with a limited set of training scenarios, the trainees simply memorize and rote learn the steps, required to be mastered to proactively solve the problems with an advent of possible accidents. In addition to that, without formal modeling, the system cannot give timely feedbacks and the performance of the trainee cannot be measured in real-time. Then to enhance the contents and usefulness of the training, the knowledge of process operations is collected and systematically stored as a form of knowledge base, in addition to the accident scenarios obtained by executing Fault Tree Analysis and HAZOP study.

By using a formal modeling language like UML and Petri nets, the accident scenarios and operators' problem-solving tasks are modeled hierarchically and classified into distinguishable subtasks. Then a new training scenario is generated on-line through automatic synthesis based on planning using the knowledge base. Safety training scenarios are generated intelligently for the tutoring of a single operator's job function as well as effective collaboration among a group of operators for real-scale malfunction diagnosis and intervention at the initial propagation stage of an accident. As case study, safety trainings for city gas governing stations are executed based on systematic scenario generation and analyzed for improved effectiveness and field applicability of the system. Reduction of time and cost in the training scenario production and real-time customized training based on new operator's level of knowledge and skill are also discussed in the framework of intelligent tutoring systems.

Experimental Study on Najas Minor of Producing Bio-Oil

Abstract: Due to the diminishing of petroleum reserves and the environmental degradation caused by consumption of fossil energy , many attentions have been paid to the application of biomass energy which is acceptable as an abundant renewable energy resource. Najas minor grows in dense clusters with highly branched stems of which a small fragment could propagate into a thrive one.In this study, we researched the properties of Najas minor to make biofuels .Firstly,the najas minor after heavy metal ions degradation were fed into pyrolyser to prepare biofuels. Significant differences of the product was observed compared with only Najas minor pyrolysed.

KEY WORDS: Najas Minor; catalytic pyrolysis; bio-oil

1.Introduction

Energy demands and environmental problems caused by fossil energy have been a matter of growing concern.Bio-oil has been considered to be a promising way for producing renewable bio-fuels or chemicals[1-3].In recent years,many countries pay attention to the importance of biomass that is an abundant renewable energy resource.Biomass can be defined as any organic material mainly consisting of carbon, hydrogen,oxygen, nitrogen, and some other components in small proportions[4].Although the biomass has been studied for nearly a century,most studies mainly concentrated in Lemna minor, Lemna gibba, Spirodela polyrhiza and Landoltia punctata.We still know little about the biological characteristics and potential application value of other varieties of biomass[5].Najas minor,a submersed herb,grows in dense clusters with highly branched stems of which a small fragment could propagate into a thrive one.Najas Minor is native to Europe, Asia and North Africa from the Netherlands to Morocco east to Japan and the Philippines,including China,Siberia, Central Asia, Iran, Turkey, Ukraine, Germany, France Italy and a host of other Countries.

Catalysts used for catalytic cracking are basically divided into the following categories: molecular sieve, alkali metal salt, metal oxide, etc.Different catalysts have different catalytic effects.The use of metal nanoparticles has shown to be very important in recent industrial applications[6].The application of precious metal nanometer materials in catalysis,which has caused the extensive concern of researchers in recent 30 years,has become a hot research topic in the field of catalysis.Gold nanoparticles have received a great deal of attentions due to their special optical and electrical performances, catalytic properties and remarkable quantum size effect, surface effect etc[7]. The synthesis of metal nanoparticles can be classified broadly into wet and dry methods[8]. The cost of production is relatively high both materially and environmentally. There is an unequivocal need to develop more cost-effective and environmental benign methods to alternate these existing methods[9].Researchers in this field turn to use organisms to produce inorganic nanostructures extracellularly[10].

Although there has been a large volume of work on the synthesis of gold nanoparticles[11-13], and obtainning bi-oil by abundant biomass, there are relatively fewer attempts to combine these two methods.In this work, gold nanoparticles were obtained by aqueous chloroauric acid solution with the of Najas Minor at room temperature. The direct pyrolysis and catalytic pyrolysis experiments of Najas Minor were carried out to observe the effects of catalysts on pyrolysis process of Najas Minor.

2. Materials and Methods

2.1. Experimental samples

The Najas Minor collected from Tianjin city and the solutions of Au(III) were used in this study.The stock solutions of Au(III) were prepared by dissolving analytical grade chloroauric acid

($AuCl_3 \cdot HCl \cdot 3H_2O$, Kewei Reagent, China) in deionized water.

2.2.Experimental apparatus and procedure

Pyrolyzer with GCMS,shown in Fig. 1,is mainly used for analyzing the composition of polymer.Pyrolyzer with GCMS includes three parts :quartz pyrolysis tube,pyrolyzer and GCMS.

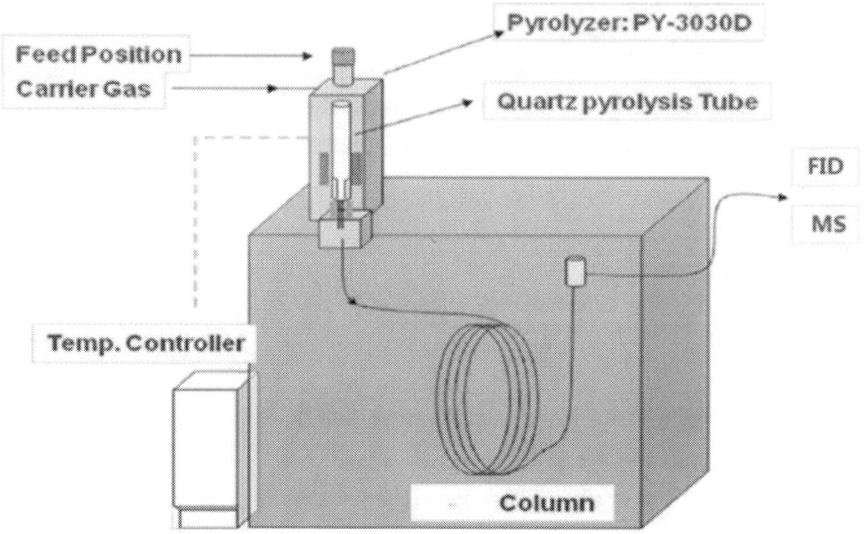

Fig. 1 Diagram of Pyrolyzer with GCMS

The quartz pyrolysis tube (shown in Fig.2) length is 12cm.The sample end diameter is 5 mm,and the pyrolysis products end diameter is 3mm.This special structure is beneficial to the fixation of the reactant in the tube.The top and bottom of the reactants are both supported by quartz wool to prevent reactants blowed by the carrier gas ,or fine particles into the chromatographic column.All pyrolysis experiments were carried out on the EGA/PY-3030D pyrolyzer(Frontier Lab,Japan) .The samples were filled in the special quartz pyrolysis tube to reduce the amount of valuable catalyst and reduce the cost of research work.

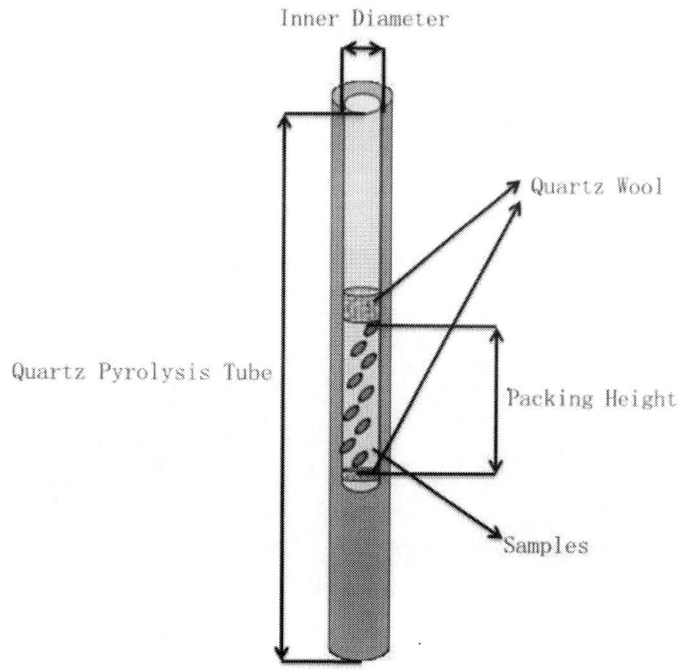

Fig. 2 Diagram of Quartz Pyrolysis Tube

Approximately 0.5g Najas Minors were used in each experiment.The samples were impregnated in 50ml deionized water and 50ml solution of chloroauric acid(0.0793mmol/L) for a night,followed by drying at 80℃ until the weight didn't change.Each experiment used 2.5mg Najas Minors as reactants.Each sample was heated, under the control of heating controller, from ambient temperature to 600 ℃ repeatly and kept at 600 ℃ for 2 h.

3. Results and discussion

3.1.MS Analysis of the Products during the Pyrolysis

Under the same pyrolysis conditions, bio-oil obtained by direct pyrolysis and catalytic pyrolysis were analyzed by GCMS,and the main chemical compounds were shown in Table1 and Table2.

Table 1 Indentification and quantification of main chemical compounds by direct pyrolysis

NO.	peak area%	Formula	Compound
1	12.67	$C_3H_6O_2$	2-Propanone, 1-hydroxy-
2	3.67	C_7H_8	Toluene
3	3.28	C_7H_8O	p-Cresol
4	3.81	$C_5H_{10}O$	Pentanal
5	3.53	C_8H_8O	Benzofuran, 2,3-dihydro-
6	2.17	C_8H_8N	Indole
7	5.93	$C_{20}H_{40}O$	3,7,11,15-Tetramethyl-2-hexadecen-1-ol
8	2.52	$C_{22}H_{42}O_2$	Phytol, acetate
9	5.25	$C_{16}H_{32}O_2$	n-Hexadecanoic acid
10	8.99	$C_{18}H_{30}O_2$	9,12,15-Octadecatrienoic acid, (Z,Z,Z)-

Table 2Indentification and quantification of main chemical compounds by catalytic pyrolysis

NO.	peak area%	Formula	Compound
1	6.01	C_7H_8O	p-Cresol
2	1.75	$C_7H_7NO_2$	3-Hydroxypyridine monoacetate
3	4.58	C_8H_7N	Indole
4	3.05	$C_{10}H9N$	1-Phenyl-1-cyclopropanecarbonitrile
5	3.90	$C_{12}H_{22}O_{11}$	.alpha.-D-Glucopyranose, 4-O-.beta.-D-galactopyranosyl-
6	8.04	$C_{20}H_{40}0$	3,7,11,15-Tetramethyl-2-hexadecen-1-ol
7	2.82	$C_{20}H_{40}$	2-Hexadecene, 3,7,11,15-tetramethyl-, [R-[R*,R*-(E)]]-
8	3.99	$C_{15}H_{28}$	1-Pentadecyne
9	2.12	$C_{15}H_{28}$	3,7,11-Trimethyl-2,4-dodecadiene
10	8.75	$C_{16}H_{32}$ 0_2	n-Hexadecanoic acid

3.2 discussing

According to the chemical category and common characteristics, the products were divided into hydrocarbons, alcohols, acids, ketones, nitrogen compounds and other compounds.Nitrogen compounds detected by GC/MS analysis ,mainly come from the pyrolysis of amino acids in Najas Minor.Table 3 showed the distribution of pyrolysis products.As shown in the table,the hydrocarbons ,alcohols and nitrogen compounds in biological oil of catalytic pyrolysis were significantly increased compared with direct pyrolysis.The content of hydrocarbons increased from 12.4% to 34.6%.The content of alcohols increased from 9.1% to 16.6%.The gold nanoparticles loaded in surface of Najas Minor, was conducive to the generation of hydrocarbon and alcohol in pyrolysis.By analyzing the GC/MS results, the main hydrocarbons in bio-oil by catalytic pyrolysis were long chain alkanes,and olefins.The main components in gasoline and diesel is hydrocarbons. As a result, whether the biomass can be effective as future power fuel oil, depends largely on the hydrocarbon content of bio-oil after further treatment.

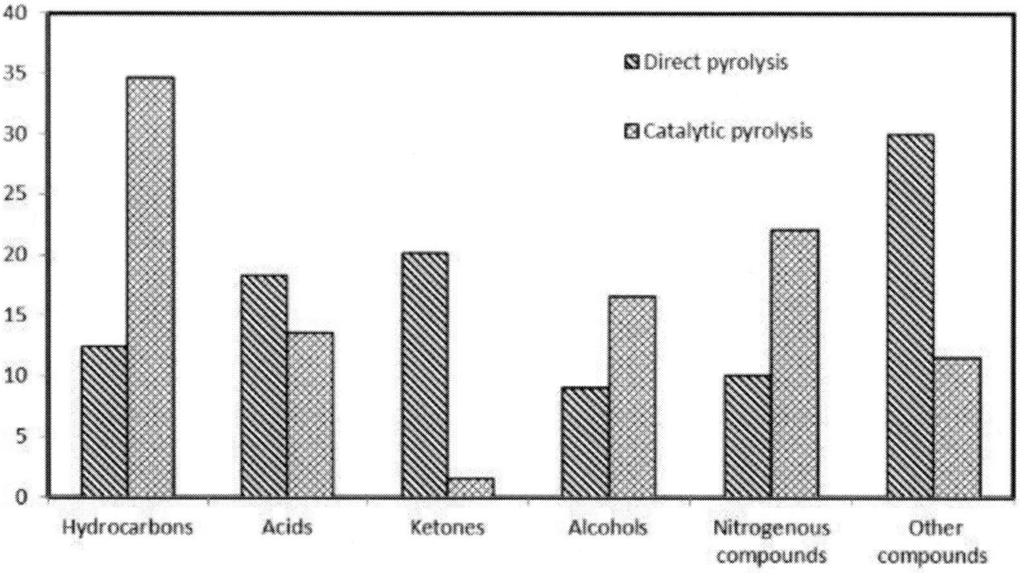

Fig.3 Distribution of pyrolysis products

4.Conclusion

Biomass was explored as a potential biomass energy resource. The experimental study on the direct and catalytic pyrolysis of Najas Minor had been presented, using Pyrolyzer with GCMS,to observe the catalytic effects of gold nanoparticles.In this study,remarkable catalytic effects for pyrolysis were observed.The pyrolysis of Najas Minor with gold nanoparticles,provided more hydrocarbons, alcohols and nitrogen compounds.

参考文献：

[1]Gong F, Yang Z, Hong C, et al. Selective conversion of bio-oil to light olefins: Controlling catalytic cracking for maximum olefins[J]. Bioresource Technology, 2011, 102(19):9247-54.

[2]Zhang Z, Bi P, Jiang P, et al. Production of gasoline fraction from bio-oil under atmospheric conditions by an integrated catalytic transformation process[J]. Energy, 2015, 90:1922-1930.

[3]Hui S, Yuqi D, Jinzhao D, et al. Transesterification of sunflower oil to biodiesel on ZrO2 supported La2O3 catalyst.[J]. Bioresource Technology, 2010, 101(3):953–958.

[4]Yaman S. Pyrolysis of biomass to produce fuels and chemical feedstocks[J]. Cheminform, 2004, 35(31):651-671.

[5]Hejný S. Ellias Landolt The family of Lemnaceae—A monographic study—Vol.1 biosystematic investigations in the family of Duckweeds (Lemnaceae) (vol. 2)[J]. Folia Geobotanica Et Phytotaxonomica, 1993, 28(1):50-50.

[6]Gardea-Torresdey J L, Tiemann K J, Gamez G, et al. Gold Nanoparticles Obtained by Bio-precipitation from Gold(III) Solutions[J]. Journal of Nanoparticle Research, 1999, 1(3):397-404.

[7]Marie-Christine D, Didier A. Gold nanoparticles: assembly, supramolecular chemistry, quantum-size-related properties, and applications toward biology, catalysis, and nanotechnology.[J]. Chemical Reviews, 2004, 35(16):293-346.

[8]Liu B ,, Xie J ,, Lee J Y, et al. Optimization of high-yield biological synthesis of single-crystalline gold nanoplates.[J]. Journal of Physical Chemistry B, 2005, 109(45):15256-63.

[9]Raveendran P, Fu J, Wallen S L. Completely "Green" Synthesis and Stabilization of Metal Nanoparticles[J]. Journal of the American Chemical Society Jacs, 2003, 125(46):págs. 13940-13941.

[10]Absar Ahmad , Satyajyoti Senapati , MIK, et al. Extracellular Biosynthesis of Monodisperse Gold Nanoparticles by a Novel Extremophilic Actinomycete, Thermomonospora sp.[J]. Langmuir, 2003, 19(8):3550-3553.

[11]Selvakannan P, Mandal S, Phadtare S, et al. Water-dispersible tryptophan-protected gold nanoparticles prepared by the spontaneous reduction of aqueous chloroaurate ions by the amino acid[J]. Journal of Colloid & Interface Science, 2004, 269(1):97-102.

[12]Gittins D I, Caruso F. Biological and Physical Applications of Water-Based Metal Nanoparticles Synthesised in Organic Solution[J]. Chemphyschem, 2002, 3(1):110－113.

[13]And A G, Murphy C J. Seed-Mediated Synthesis of Gold Nanorods: Role of the Size and Nature of the Seed[J]. Chemistry of Materials, 2004, 16(19):3633-3640.

COMPARATIVE ANALYSIS OF MOLECULAR STRUCTURE IDENTIFIABILTY BASED ON SIGNATURES AND DESCRIPTORS

Zelimir F. Kurtanjek
University of Zagreb, Croatia

Abstract

This work is focused on molecule computer representation aimed for molecule predictive model (QSAR) and inverse modelling. Two different approaches are compared, i.e. the molecule structure coding based on graph theory (Faulon et al.[1] extended valence) and the chemical based descriptors[2]. Applied are 1-3rd level molecule graph fragments and 630 2D molecular descriptors. The molecule test sets are alkanes, alkenes, acetones, aromatics, organic acids and halogenated hydrocarbons in the range of C_1-C_{12}, selection of biomolecules (including aminoacids) and a set of binary ionic liquids (cations: imidazole, pyridinium, quinolinium, ammonium, phosphonium). The comparative comparison of molecule description based prediction potential are statistically evaluated by octanol-water (PWA) partition coefficients (for hydrocarbons), viscosity of ionic liquids and biological impact on acetycholinesterase inhibition. The evaluation is based on two distinct levels. On the first level the comparative analysis based on internal and external correlation between "scores" in multidimensional spaces of the separate first, second and third order valance spaces, and the molecular signature space. Firstly, the potential of information content applicable for inverse modeling is tested on ability for molecule classification by use of linear discrimination (LDA) based on the singular value decomposition (SVD) of corresponding molecular fragments and descriptors. For chemical-biological property inverse modelling proposed is application of constrained genetic algorithm (GA). Tested are linear structure multivariate (chemometric) models such as principal component regression and partial least squares and nonlinear decision trees (random forest). Applied are singular value decompositions for reduction of the molecule signature and the descriptor spaces up to the significance level of p=0.05. The selected level of significance is presumed to be in coherence with experimental data of molecule physical and chemical properties. The main focus is on extraction of the main molecular fragments (partial graphs extracted from molecular signatures) and specific molecule descriptors. Importance of application of open source software modules provided by Bioclipse[3] and R software packeges[4] for computational-statistical numerical algorithms is emphasized.

Key words: Molecule signatures, descriptors, inverse modelling, decision tree

Introduction

Advancement of high-throughput experiments have lead to „big data" databases applied in pharmaceuticals, medicine, new material science, catalysis, and new industrial processes. The role of chemoinformatics as a science for computer encoding of molecule properties and algorithms for data mining and inference are the essential steps in the development. Computer encoding of molecular properties has been a part of chemical engineering from the beginning of process design. It is mostly based on molecule group contributions, but has been extended by application of molecular descriptors (1-2-3 D), molecule signatures, and application of molecular quantum mechanics. Modeling is faced with the problem of „high dimension curse". For example, about 5 000 various molecular descriptors can be determined for each molecule, although most of them are highly correlated. It results in an ill-conditioned problem of modelling with large uncertainties in parameter and poor model prediction accuracy outside modelling training sets. Great improvement has been achieved by application by use of model ensembles, like random forest of many individual decision trees. These models are highly flexible, can be trained on discontinuous and nonlinear functional relations, can encompass various data formats like numerical and categorical variables, and due to large number of randomly selected individual predictors in principle do not have tendency to overfitting. However, when the object of modelling is extraction of important molecular features, due to autocorrelation between descriptors random forest, produced lists of importance are large sets of poorly distinguishable variables. The focus of this work is to apply feature regularization for optimization and reduction of list of variable importance. Compared are two methods for structure encoding, molecular descriptors and molecular signatures of first and second order.

Modeling and Factor Regularization

In this work are selected two distinct data sets of organic molecules from open source data bases (J. Sangster, 1989) and ionic liquids ILs (The UFT/ Merck and NIST, 2015), and lists of the selected molecules are given in Table 1. As target properties for organic molecules are selected the corresponding molecular classes and partition octanol-water (POW) coefficients, and dynamic viscosity in the range of 30-80 ^{0}C of ILs is selected The molecules are coded into SMILES and used as input lists into software for evaluation of molecular signatures. Most popular open source software tools are DRAGON (R. Todeschini, et al., 2009) and PADEL (Chun Wei Yap, 2010). Here are used 1-2 D descriptors generated by PADEL. For the same list of molecules are determined first and second order molecular signatures or extended valances (J.P. Faulon, 2003). The signatures are two dimensional graph based structural molecular fragments. The vertices are atoms and the edges are corresponding bonds (lengths or heights), and are here only used of the first and second level. A molecule is characterized by a complete set of signatures as input data for modelling. Since the models are based on large number of variables (molecular features) possibly lager than number of samples (molecules), usually denoted as p>>n models. They are intrinsically „ill conditioned" leading to good predictions on train data but very poor predictions on test data. Also, produced list of variable importance are very dense "clusters are not distinguished" with large uncertainties (broad confidence intervals). This makes the problem of inverse modelling aimed to predict a set of candidate molecules for given set of target properties, such as in a case of high dimension combinatorial task for selection of ILs cations and anions, very unreliable. In order to reduce model dimensionality and improve robustness and parsimonious of important molecular features here is applied modelling with feature regularization. The model is based on ensemble of random trees (CART models, L. Breiman et al. 1998) with included feature regularization.

Data sets of organic molecules	**Range of number of atoms**				
Molecular classes	C	H	O	Cl	F – I - Br
Alkanes	5-12	12-30			
Alkenes	4-9	6-18			
Acetones	6-24	6-26			
Halogenated hydrocarbons	3-10	6-22	1		
Organic acids	1-16	2-32	2		
Esters	4-11	6-22	2		
Aromatics	1-12	1-21		3	2 – 2 - 1
Data sets of ionic liquids (ILs)					
Cations	**Anions**				
1,3-Dimethylimidazolium (+)	PF_6 (-) Cl (-) CH_3COO (-)				
1-Methylpyridinium (+)	BF_4 (-) $MeSO_4$ (-) $EtSO_4$ (-)				
1,1-Dimethylpyrrolidinium (+)	Tf_2N (-) CF_3SO_3 (-)				

Table 1. The molecular groups of the organic and ionic liquid molecules selected for random forest models and extraction of the most important molecular features. The properties for the organics are obtained from J. Sangster (1989) and for the ionic liquids from Merck and NIST data bases (2015). Molecular descriptors and signatures of ILs ions are recorded separately and adjoined as a single record.

The modelling is formulated as an optimization (functional and multivariable minimization) problem with two parts of the objective function (T. Hastie et al., 2009).

$$\min F_{object.}(X, T, \omega) = SSE(y, \hat{y}) + \phi(T, \omega) \tag{1}$$

The first part is the sum of square errors (SSE) based on difference between observed y and prediction $\hat{y}$ based on input data by an ensemble T of interrelated CART models (L. Breiman et al. 1998), The second part is a selected metrics ϕ for the model complexity depending the sum of leaf probabilities for population of individual CART models and their total number N_T.

$$\phi(T, \omega) = \gamma \times N_T + \frac{1}{2} \times \lambda \times \sum_i^{N_T} \omega_i^2 \tag{2}$$

The ensemble output prediction is a sum of predictions of individual T_i tree CART models:

$$\hat{y} = \sum_i^{N_T} T_i(X) \tag{3}$$

A very efficient open source code Extreme Gradient Boosting XGB (T. Chen, 2016) is available in R language (R Developm-ent Core Team, v. 3.2.3, 2015). The optimization algorithm is consecutive, starting with first CART model which is appended with a next tree boosted by errors in predictions of previous tree. At each step optimization is solved by linearization of the objective function to quadratic terms. Number and structure of trees as well as the parameters for regularization need to be found by trial and error and minimal number of trees by folding and cross validation error on the training set. The critical step is validation on untrained set to have acceptable model error and avoid model overfitting with minimal model complexity.

Results and Discussion

Tested are models for prediction of the seven molecular classes (Table 1.) based on molecular descriptors and molecular signatures (first and second level). Results for molecular descriptor feature importance are shown on Fig. 1. The effect of regularization is very pronounced. Out of hundreds of descriptors for the molecular classification needed are only few, the most important being BCUTw.1h. On contrary, when an ensemble of independent CART trees is applied, i.e. random forest model, level of accuracy of prediction is the same, but the list of variable importance is densely populated with numerous descriptors.

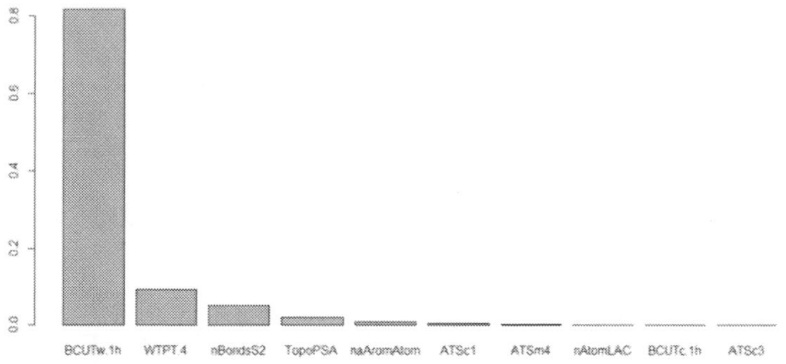

Fig 1. Variable importance molecular descriptor factors for the inverse modeling of the selected molecular classes. The first five descriptors are: BCUTw.1h (based on a weighted version of the Burden matrix which takes into account both the connectivity as well as atomic properties of a molecule), WTPT.4 (Randić's weighted path), nBondS2 (bond count), and naAromAtom (number of aromatic atoms).

The same classification problem was tested by the first and second order molecular signatures. The important signatures are depicted on Fig. 2-3. Accuracy of prediction on trained sets of molecules compared to descriptors is about the same, though the importance factors are not as distinctive as in case of the descriptors.

significantly better when molecular descriptors are applied. The

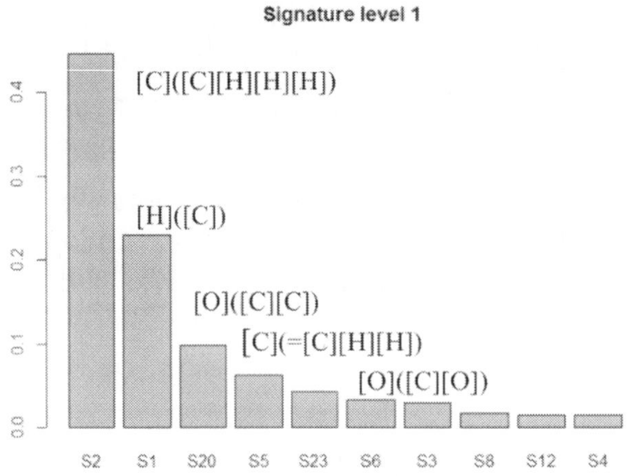

Fig 2. Relative importance of the individual first level signatures for inverse prediction of the selected molecular classes. The first five most important signatures are coded according to SMILES (Simplified Molecular Input Line Entry System).

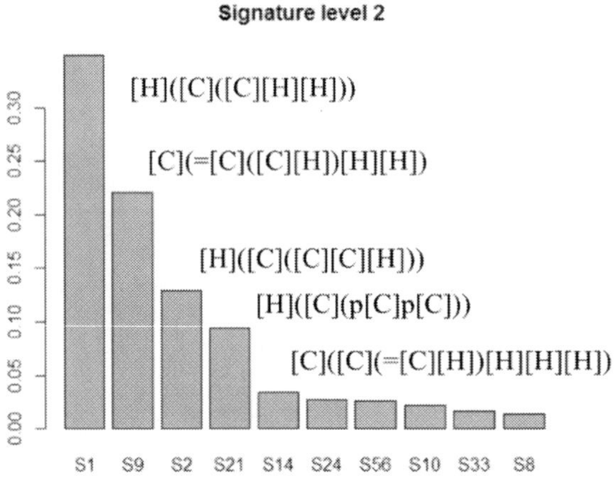

Fig 3. Relative importance of the individual second level signatures for inverse prediction for the selected molecular classes. The first five most important signatures are coded according to SMILES (Simplified Molecular Input Line Entry System).

However, model accuracy on untrained set of molecules, depicted in Fig. 4, is significantly lower for the model with molecular descriptors compared to signatures. The average error with descriptors is about 5% while for signatures about 11%. The "box and whiskers" plots are obtained by error evaluation of ten randomly generated untrained sets.

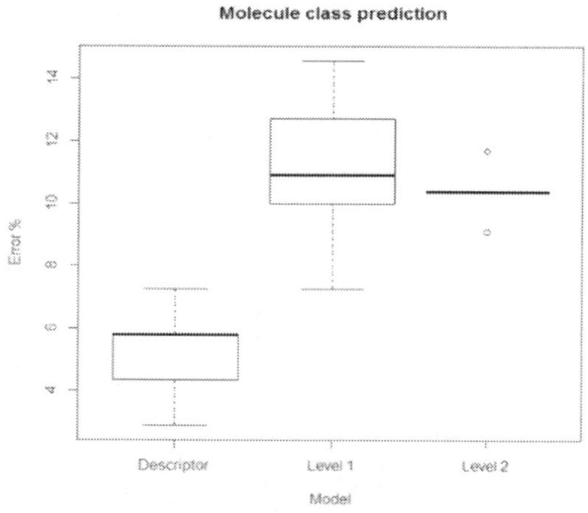

Fig 4. "Box and whiskers" plot for the model validation (relative errors) for inverse prediction of the molecular classes based on the set of untrained molecular descriptors and the first and second level of molecular signatures.

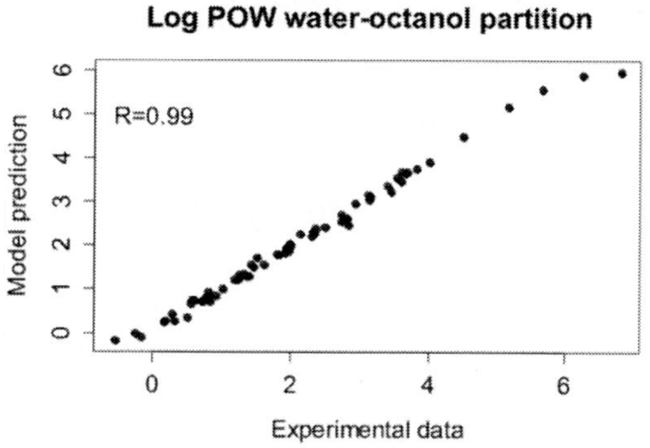

Fig 5. Predictions of the partition octanol-water coefficients POW, for the specified molecular classes of organic molecules, are based on the random forest model with the molecular descriptors.

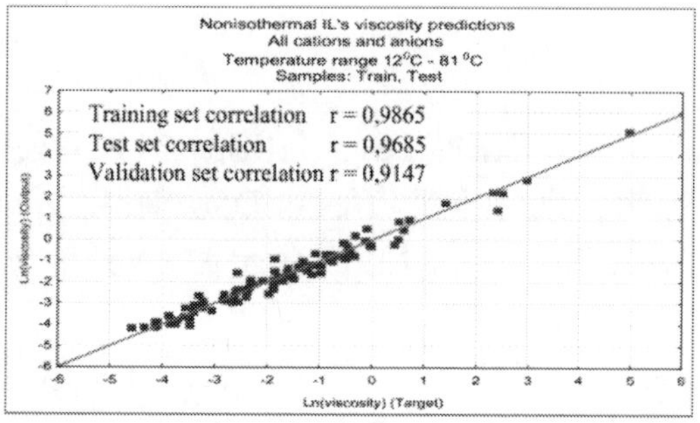

Fig 6. Predictions of dynamic viscosities for the specified groups of ionic liquids (ILs) based on the random forest model with the molecular descriptors. The viscosities cover the temperature range from 12 – 81 ^{0}C and five orders of magnitude.

In Fig. 4 can be observed that dispersion of errors (quantiles) for the second level molecular signatures is negligible, though two outliers are present.

For prediction of molecular properties applied are random forest models based on molecular descriptors. For all of the organic molecules, included in the seven classes, evaluated are octanol-water partition coefficients POW, shown in Fig. 5. The range of POW values is very high, on log scale fro 0 to 6, but the obtained accuracy of the model predictions is high (Pearson correlation R=0.99). The same approach is applied for prediction of ionic liquid viscosity data in temperature range from 12 to 81 ^{0}C. Since temperature effects on viscosity are intrinsically nonlinear, they represent a modelling challenge. However, the decision trees are very efficient for accounting nonlinearities, though nonlinearity is reflected in non-orthogonal decision planes, rendering trees with significantly increased number of branches. For the range of 6 order of magnitude the accuracy of prediction on trained set is R = 0.98 and on untrained set R = 0.91.

Conclusions

Compared are ensemble decision trees models for molecular feature class and property prediction based on extensive list of molecular descriptors and first and second level of extended valances or signatures.

The aim was to test effects of regularization on ensemble of CART decision tree models. Proven is effectiveness of the extreme gradient boosting algorithm in cases with extensive list of (molecular) features.

Obtained is increased accuracy on untrained sets of data, and more importantly, obtained are parsimonious lists for ranking of variable importance lists. This enables further significant reduction in development of robust and simple decision tree models.

The reduction in number of molecular features applied for decision tree models may greatly simplify the challenging problem of inverse modeling based on molecular descriptors and/or signatures.

References

1. J. Sangster, J. Phys. Chem. Ref. Data, **18**(3)(1989) 1111-1227
2. The UFT/ Merck Ionic Liquids Biological Effects Database, www.il-eco.uft.uni-bremen.de
3. NIST, Ionic Liquids Database Standard Reference Database #147, <ilthermo.bouder.-nist.gov /ILThermo/mainmenu.uix>
4. R. Todeschini and V.Consonni: "Molecular Descriptors for Chemoinformatics", (2 volumes), WILEY-VCH,Weinheim (Germany) 2009.
5. Chun Wei Yap, PaDEL-Descriptor: An Open Source Software to Calculate Molecular Descriptors and Fingerprints, J. Comp. Chem., (2010):**32**: 1466-1474.

6. J.-L. Faulon, D.P. Visco Jr., R.S. Pophale, The signature molecular descriptor. 1. Extended valence sequences vs. topological indices in QSAR and QSPR studies, J. Chem. Inf. Comput. Sci. **43** (2003) 707–720.
7. J.-L. Faulon, C.J. Churchwell, J.D.P. Visco, The signature molecular descriptor. 2. Enumerating molecules from their extended valence sequence, J. Chem. Inf. Comput. Sci. **43** (2003) 721–734.
8. L. Breiman, J. H. Friedman, R. A. Olshen, C.J. Stone, „Classification and Regression Trees", CRC, Boca Raton, 1998.
9. T. Hastie, R. Tibshirani, J. Friedman, The Elements of Statistical Learning, Data Mining, Inference and Prediction, Springer Verlag, Second Edition, New York, 2009.
10. T. Chen, T. He and M. Benesty (2016). xgboost: Extreme Gradient Boosting. R package version 0.4-3. https://CRAN.R-project.org/package=xgboost
11. R Development Core Team, , R: A language and environment for statistical computing. R, Vienna, Austria,www.R-project.org 2015.

444478 Biogas Reforming with Addition of Carbon Dioxide to Syngas over Ni-Based Bimodal Pore Catalyst

Abstract:

Methane dry reforming of (MDR) has been widely studied due to its important applications in producing of syngas ($CO + H_2$), removal of two greenhouse gases, and upgrading of biogas (mainly composed of CH_2 and CO_2) into value added chemicals. Compared to the noble metal catalysts, nickel-based catalysts are suitable for industrial scaled-up MDR due to their high activity and low cost. However, nickel-based catalysts have been encountering a significant challenge that is the active metal sintering and coke deposition, leading to the catalyst deactivation and then the poor stability. Nevertheless, it has been reported that the deactivation of nickel catalysts can be suppressed by adding promoters like strong Lewis bases with the enhancement of chemisorb CO_2, such as MgO and CaO, and like lanthanide elements with the capacity of oxygen storage and release, such as CeO_2 and La_2O_3. Al_2O_3 is a commonly applied catalyst carrier own to its good pore size dispersion, high specific surface area and high mechanic strength.

Herein, we report that the bimodal NiCeMgAl catalysts with different NiO-loading were synthesized by the refluxed co-precipitation method and were evaluated for MDR performance by using real biogas from anaerobic digestion of food wastes. The sample containing 15wt% NiO-loading (Ni_{15}CeMgAl) was found to be active enough at 750 °C with a high CH_4 conversion of around 90%. Ni_{15}CeMgAl catalysts kept their bimodal porosity after reduction and MDR process. The evolution of the Ni_{15}CeMgAl catalyst before and after MDR process was investigated by SEM, TEM, XRD, and TGA techniques. The correlation between the structure evolution and catalytic performance change was also discussed.

444994 Multiscale Simulation of Proteins Motion in the Microchannel

Abstract:

Protein adsorption is an important phenomenon, which plays a prominent role in chemical, industrial and medical areas. In this work, the dynamic behavior of the egg-white lysozyme has been investigated in a micro-channel with a multiscale modeling approach. The Molecular Dynamic simulation, free energy computation and Langevion Dynamics (LD) simulations have been combined to study lysozyme adsorption process on Au (111) surface. The gold surface has strong attraction to proteins. The protein-surface adsorption or desorption free energies is calculated. From the case studies of LD simulations, the surface has effects on the protein dynamic adsorption process. By the comparison of LD, MD and the experiments, the time step is concluded as an important factor to study protein adsorption behavior at the water-surface interface. Fine time step at the scale of 1 picosecond is necessary to provide enough accuracy to resolve proteins' dynamics trajectories.

444827 Syngas to Ethylene Glycol: A Technology Overview

Abstract:

SABIC is the world largest MEG producer petrochemical company and most of its MEG is being produced through the hydrolysis of EO. Ethylene oxide (EO) is a simplest form of epoxides and due to its molecular structure, acts as a versatile intermediate for a variety of chemicals such as; monoethylene glycol (MEG), di- and triethylene glycols, ethoxylates, ethanol amines, and glycol ethers. The growing demand of MEG owing to its downstream utilization in polyester industry has prompted the research of an efficient, cost effective, and environmentally benign EO production process.

The major feedstock for MEG are ethylene and syngas. The heavy dependence of MEG producers on ethylene and its scarcity in some parts of the world promoted the need of alternate feedstock for highly demanded MEG. The technological developments utilizing syngas as a feedstock will be discussed along with the commercial analysis and future growth in EO/EG business.

445029 Electrospun Nanocomposite Fibers for the Removal of Heavy Metals in Water

Abstract:

Uniform electrospun nanofibers with small diameters were successfully prepared by systematically optimizing the electrospinning parameters including polymer concentration, salt loading, applied voltage, tip-to-collector distance, and flow rate. These parameters are shown to affect the fiber morphology and the addition of iron (III) nitrate salt was found to be able to reduce the fiber size. Due to the conductive nature of iron salt in polymer solution, even a relative lower voltage of 6 kV could generate a stable fluid jet and yield the nanofibers. The thermal properties of the composite fibers were investigated by thermal gravimetric analysis (TGA) and differential scanning calorimetry (DSC). Rheological studies were also conducted. The composite nanofibers were further carbonized to carbon nanofibers. Applications of the nanofibers in environment remediation have shown promise.

445004 Grand Challenges in Future Energy and Chemicals: Climate Change and Sustainability As Drivers for New Technology and Innovation

Abstract:

Global demand for energy will double by 2050, driven by increasing population and affluence especially in emerging economies. While fossil fuels will continue to underpin the energy supply for the near future, concern over impact on climate change continues to grow, driving interest in expanded use of renewables and development of carbon capture and sequestration or re-use solutions. In addition to CO_2, water use and air emissions are critical issues to be addressed in future and current energy supply, as is energy storage for effective use of renewables. This presentation will provide an overview of what industry is doing today to address these issues, including the factors and metrics used in decision making.

444825 Significance of Energy Intensity and GHG Emission Reduction Towards Sustainability of a Petrochemical Plant

Abstract:

Sustainability is an important pillar in today's petrochemical industry for improving process efficacy by reducing specific energy consumption, greenhouse gas (GHG) emissions, and wastewater. SABIC (Saudi Basic Industries Corporation) like other giant petrochemical companies has set stringent sustainability targets and investing huge resources to transform its plants from the present position to the first quartile in each category.

SABIC is the world largest producer of ethylene oxide and ethylene glycol. For each ton of glycol produced, 0.2 to 0.5 tons of CO_2 is emitted as GHG. The significance of sustainability "in general" while the methodology adopted to reduce energy intensity, carbon footprint, and wastewater "in particular", will be discussed thoroughly in our talk.

440527 Recent Developments and Current Insights in Ethylene Flare Technology: The Impact of Burner Design, Vent Gas Variability and Ancillary Equipment

Abstract:

The use of physical properties of flare vent gas and the exit velocity have commonly been used to describe flame stability for flares. Multipoint ground flare burners, or pressure assisted flare burners, commonly used in the ethylene industry do not abide by the EPA mandated exit velocity criteria.

A review of available data reveals that neither combustion zone net heating value nor combustion zone flammability limit are good predictors of combustion efficiency for pressure-assisted flares. Statistics calculated from the data also indicate that the manufacture of the burner head is also significant to determining the combustion efficiency of the flare although all of the data available showed the burners outperformed current EPA requirements.

A new burner design with advantageous flame characteristics is presented along with flame stability and combustion efficiency data from physical testing for multiple burner types. Insight is gained thought computational fluid dynamic simulation of the designs.

A discussion of the practical implications of ancillary equipment, such as knock-out drums, will also be provided.

444797 A Reduced Combustion Mechanism for the Simulation of Flaring

Abstract:

Flare is widely used in energy and petrochemical industries to burn out unwanted combustible gases. The detailed combustion mechanisms are too complicated for the speciation study of flaring events with Computational Fluid Dynamics. In addition, the more rigorous ANSYS FLUENT Eddy-Dissipation Concept (EDC) model, which describes turbulence-chemistry interaction, can handle a maximum of only 50 species. To alleviate this computational difficulty, a less than 50 species reduced mechanism is needed. In order to predict the generation of soot and VOCs during the combustion of C1-C4 hydrocarbons accurately and efficiently, a reduced combustion mechanism was generated, based on the detail mechanism in the literature. This approach consists of the generation of skeletal mechanisms using directed relation graph with specified accuracy requirement. The subsequent generation of reduced mechanisms uses computational singular perturbation based on the assumption of quasi-steady-state species. The reduced mechanism was further validated successfully against experimental performance indicators like ignition delay time and adiabatic flame temperature using CHEMKIN. This Reduced mechanism can be used in ANSYS Fluent to predict the emissions and DRE, CE of hydrocarbon flares.

443490 Modeling Flare Performance Using Closed-Form and B-Spline Mathematical Functions

Abstract:

Many industries seek to avoid incompliance with the current/upcoming EPA regulations for flare performance and meet the stipulated flare efficiencies along with a smoke-less flaring. To achieve the target performance along with satisfying the EPA regulations, it is necessary to understand the influence of various operating parameters on the overall process. Studying the effect of operating parameters through experiments is tedious, expensive, and time consuming. It would be more cost-effective and organized to use validated models to guide flare operations. With this objective, mathematical models have been developed to estimate Combustion Efficiency (CE), Destruction Efficiency(DE) and Opacity over a wide range of operating conditions for both steam and air assisted flares . For this purpose, the flare test data with soot emissions for the years 1983 to 2014 (including 1983/1984 EPA, 2010 2010 TCEQ tests, and 2014 Carleton University) were used. Multi-variable parameterized sigmoid /statistical distribution models were developed for the data involving various types of fuels like propylene, propane, natural gas, methane and ethylene. The variables like vent gas Net Heating value at Combustion Zone (NHVcz), Lower flammability limit at Combustion Zone (LFLcz), Combustible concentration at combustion Zone (Ccz) were calculated and used as inputs along with exit velocity(v) for the parametric models . The study was conducted using various MATLAB® toolboxes.The developed models were found be in good agreement with the experimental data with a goodness of fit of above 0.9 for both the cases of CE and DE. For the same data, B-form Cubic spline surface modeling has also been done and the results are presented for comparison. The robust models with high fidelity are envisioned to be used to find conditions that meet specified flare efficiency/opacity and that optimize flare operations in economic terms.

16AIChE
Spring Meeting, Houston, TX

Calculation of Flare Header Pipe Temperatures during Depressurizing

Philip Diwakar
Bechtel Oil, Gas & Chemicals, Inc.
3000 Post Oak Blvd., Houston, TX

Jaleel Valappil
Bechtel Oil, Gas & Chemicals, Inc.
3000 Post Oak Blvd., Houston, TX

Prepared for Presentation at
American Institute of Chemical Engineers
2016 Spring Meeting
11th Global Congress on Process Safety
Houston, Texas
April 10-14, 2016

UNPUBLISHED

Calculation of Flare Header Pipe Temperatures during Depressurizing

Philip Diwakar
Bechtel Oil, Gas & Chemicals, Inc.
3000 Post Oak Blvd., Houston, TX

Jaleel Valappil
Bechtel Oil, Gas & Chemicals, Inc.
3000 Post Oak Blvd., Houston, TX

Keywords: Computational Methods, flare system, LTCS, risk analysis

Abstract

LNG facilities have wet and dry flare systems that need to handle cold relief flows for short time periods. The wet flare header lines are typically carbon steel while most of the dry flare header lines are stainless steel. However, the use of stainless throughout the flare system is very expensive especially if there are lines leading from one train to another. It is customary to use stainless steel in extreme low temperature lines and transition to impact or non-impact tested carbon steel after the temperature in the lines have been established to be higher than the minimum tested temperatures for Low Temperature Carbon Steel (LTCS). This paper discusses the calculations required to evaluate the use of LTCS piping in cold intermittent service on dry flare headers and the associated stresses due to differential thermal expansion of the materials in the event of a risk analysis review by process safety. Failures lead to loss of containment, possible fire, impact to personnel and the environment. Transient and steady state heat transfer calculations are used to estimate the distance over which the low temperature relief flows impact the piping and assess the suitability of transitioning to the LTCS material.

1. Introduction

In a gas processing facility temperatures below -25°C are common. Even warmer streams such as natural gas liquids (NGL) products or the higher-pressure gas streams have the potential to go below -25°C during pressure letdown due to Joule-Thomson cooling. Operating standard carbon steel piping at temperatures below -25°C can lead to cold brittle fracture failure of the piping since such piping has a low temperature allowable limit of -29°C (-20°F). Charpy V notch impact tests are typically used for toughness testing of piping components to meet ASME B31.3 standards. Newer low temperature, impact tested carbon steel alloys can withstand temperatures as low as -60°C without failure but these are almost as expensive as stainless steel. The purpose of this paper is to present calculation methods to obtain the length of stagnant gas piping for which the temperatures are below -29°C to help designers select the appropriate material for the piping.

From a stress point, the B31.3 criteria is as follows. Although the minimum design temperature for nonimpact-tested carbon steel piping with a wall thickness less than 0.5 in. is -29°C, B31.3 allows for "occasional variations of pressure and/or temperature which shall be considered in selecting design pressure and design temperature". The criteria include the following

a) The design temperature is at or above -46°C.

b) The maximum operating pressure of the manufactured components will not exceed 25% of the maximum allowable design pressure at ambient temperature.

c) Longitudinal stress due to pressure, dead weight, and displacement strain does not exceed 6ksi (41 MPa).

An addendum to B31.3 issued in 2000 modified the requirements and are similar to the fracture mechanics guidelines found in the pressure vessel code ASME Section VIII. The addendum is less restrictive in the application of non-impact tested carbon steel and allows for intermittent temperatures lower than -46C.

The Australian standard AS2885.1 has a specific requirement dealing with these short term low temperature excursions and these are related to the hoop stress in the piping. For temperatures lower than -29°C, the hoop stress may not exceed 85 MPa for any piping component at any time. As an example, for a feed gas pipeline, pigging operations are limited at least 10% below this design margin so as not to exceed minimum design metal temperature (MDMT).

Minimum temperatures during pressuring a slug catcher for example may be studied using Computational Fluid Dynamics (CFD) and process simulation tools (HYSYS, VMGSim or other tools) to determine if the metal wall temperatures can exceed the piping MDMT. In this paper we site two examples for analyzing low temperature excursions and related metal temperatures.

2. Pressurizing and low temperature excursions at branch connections

Our first example deals with pressurization of lines downstream of the main inlet feed isolation valve. Feed gas at high pressure is used to pressurize a large downstream volume through a large header (typically 42 to 60inch diameter). During startup a small bypass (typically 2-4inch diameter) is open and flow is reduced to a choked flow as it passes through a valve before entering the larger line. Because of sudden expansion, there is a Joule-Thomson cooling effect at the tee connection between the header and branch. Temperatures are likely to drop below the allowable LTCS temperature. For this analysis, the opening of the valve over a finite period of time is considered as high pressure gas expands into the main header with subsequent temperature drop that is convectively transferred to the header and the tee joint at the bypass. The boundary conditions at the valve exit which is the inlet for the CFD simulation - the mass flow, pressure and temperature - are derived from the VMGSim tool for the choked flow conditions through the smaller bypass system. Figure 1 shows the schematic of the system used for this study.

An external heat transfer convective boundary condition is applied to the outer wall of the main header pipe and the bypass pipe with a free stream ambient temperature. In order to capture the gas expanding into the main header and into a large volume on the downstream side, a sphere with radius representative of the pressurizing volume may be modeled instead of modeling the entire downstream piping system. Gas expands and pressurizes the volume over a long period of time. At the inlet (exit of valve), a mass flow inlet boundary condition may be used in lieu of pressure. The mass flow is increased in stages to account for valve opening with the temperature equal to the exit temperatures from VMGSim over the period of valve opening. This may involve a transient multiphase calculation depending on initial gas composition. An evaporation-condensation utility is used to capture the throttling process (Joule-Thomson cooling effects) and capture any liquid formation occurring during sudden expansion. Pressure and temperature is monitored on the walls, bulk of fluid and in the shock region over time at the incoming bypass line and in the header pipe.

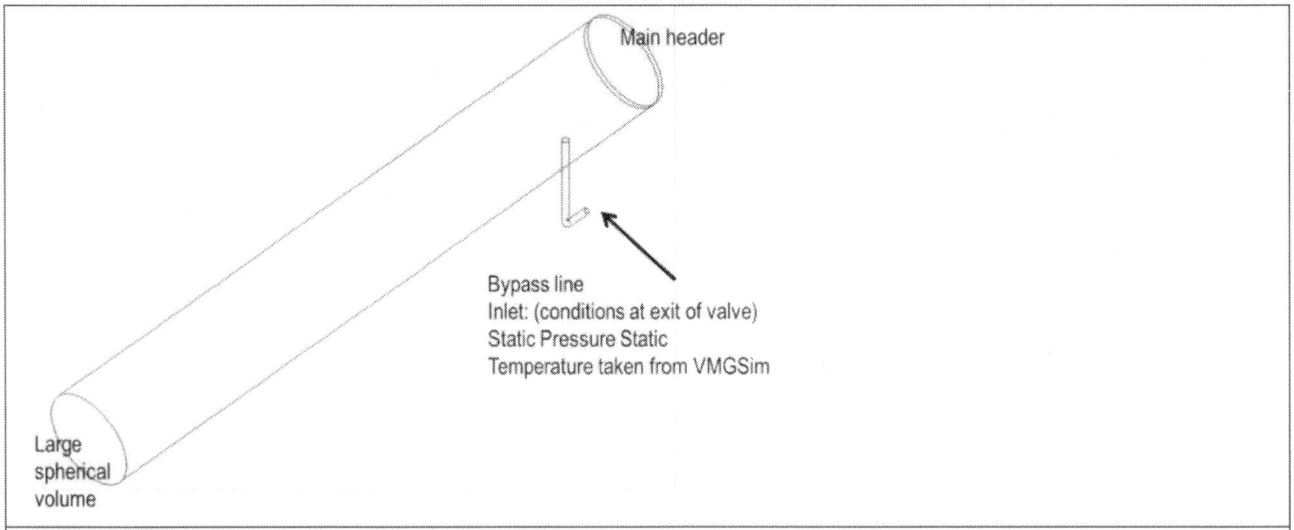

Figure 1: Depressuring line setup CFD model

Figure 2 shows temperature profiles scaled from -50°C to 7°C at the walls of the bypass line and main header at the branch connection. It can be seen from the temperature profiles that some degree of flashing and subsequent choking occurs in the smaller line itself and the maximum Joule-Thompson cooling occurs on entry into the larger space of the header line.

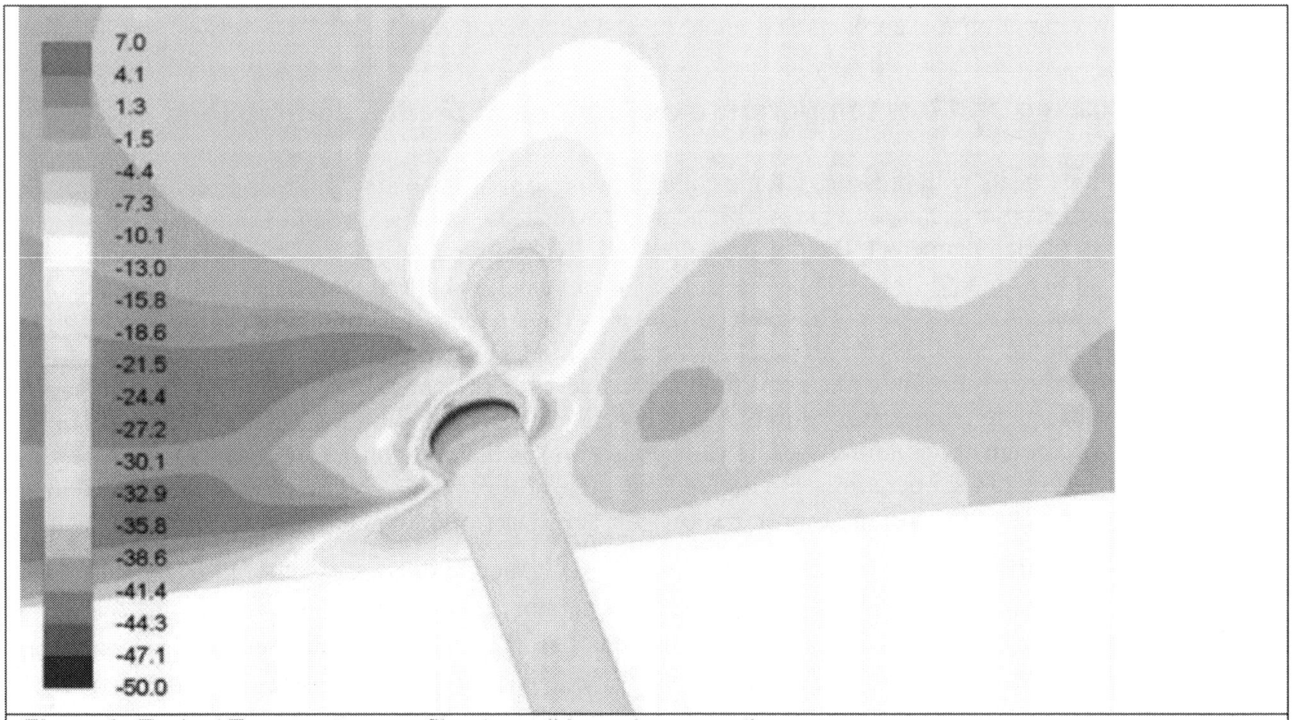

Figure 2: Typical Temperature profile at small branch connection

Liquid flashing to vapor in the main line can be observed looking at the density contours as shown in Figure 3. Choking and flashing in the smaller bypass line are also shown as density contours in Figure 4. The temperature at various locations, including at the branch connection and the opposite side where the jet impinges may be monitored to determine how low the metal temperature may reach over the time of pressurizing and valve opening. A typical temperature curve is shown in Figure 5. The analysis shows how low the metal temperature can go below and for how long. These results may then be supplied to the materials specialists to assess the suitability of LTCS around the small bore connection and surrounding regions. The pressurizing sequence and the valve opening time may be then adjusted to obtain a benign temperature spike to suit the material of piping to withstand brittle fracture due to impact loads or other factors. In general, this localized area of low temperatures has to be stainless steel to avoid a very long depressurizing schedule extending several weeks. Otherwise specialty fittings and a fracture prevention plan should be proposed leading to additional cost.

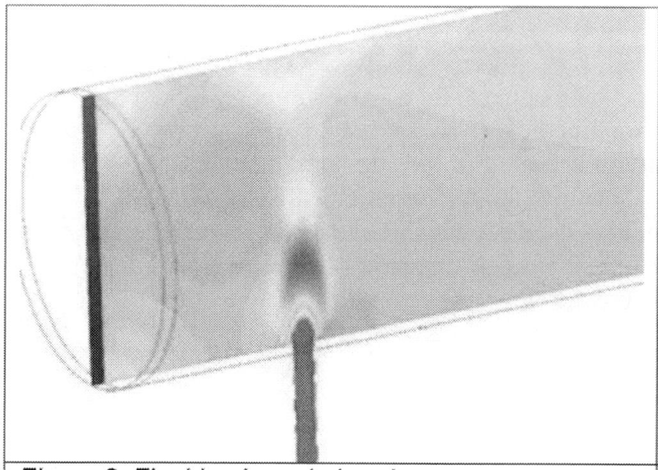

Figure 3: Flashing in main header

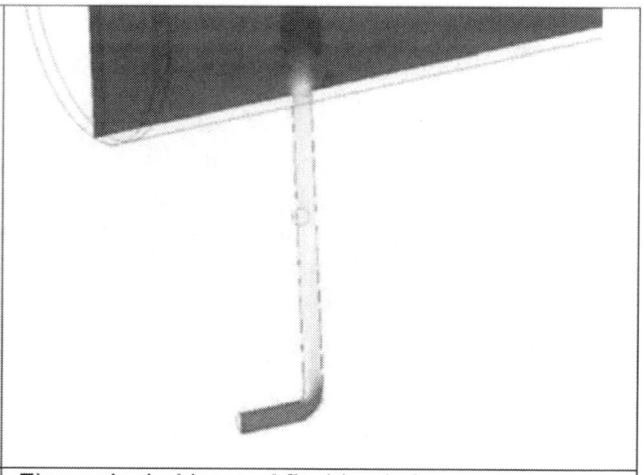

Figure 4: choking and flashing in bypass

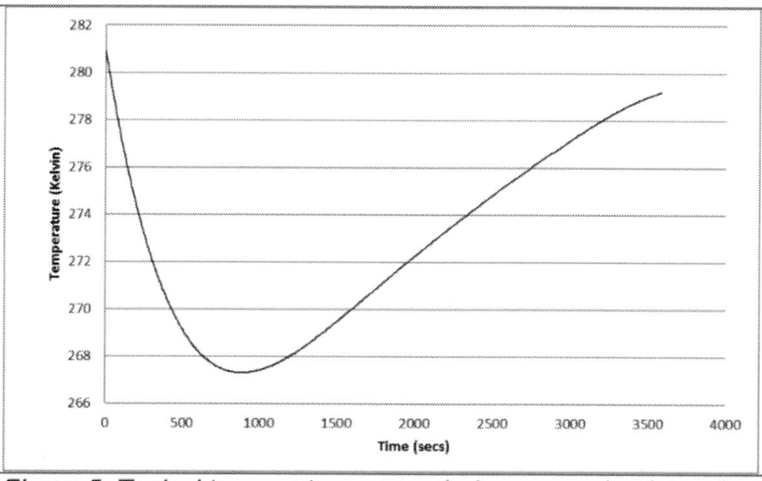

Figure 5: Typical temperature curve during expansion into a large header.

3. Flare header stagnant line temperature determination during depressurizing operation

As a second example for low temperature excursions, a CFD calculation is used to determine the temperature profile and the distance along several stagnant gas pipe laterals of different diameters extending from the main wet flare header where the inner pipe wall temperatures are less than -29°C during trunk line depressurizing. This exercise is to determine what length of impact tested LTCS is to be used for each diameter lateral in place of standard carbon steel. A schematic of the main header with several diameters of branched piping is shown in Figure 6. The proximity of a lateral to the depressurizing valve has some influence on the temperature distribution on the inner wall of the pipe.

However, in general, the variation in temperature appears to be linear after a long period of time (steady state). Since the temperatures in the main header are usually below any allowable for LTCS material, these are stainless steel. The CFD calculation determines how far along the branches, the stainless steel can be replaced by carbon steel.

The model used for the CFD calculation is shown in Figure 6. The flare header line from adjacent trains as well as other smaller diameter lines, are joined to the wet flare header with butt-welds for the purpose of this analysis. Wall thickness for metal mass is modeled. For convective heat transfer from fluid to the wall and through-wall conduction a higher resolution boundary layer as shown in Figure 7 is constructed.

For the CFD calculation, linear variation of gas properties is assumed. To avoid constructing a long length of pipe, the stagnant volume is represented by a hemisphere (for a half-symmetry model). Conservatively, pipe bends and elbows along the main pipe header are ignored, since the pressure drop across fittings and equipment will increase the gas temperature. For the smaller diameter pipes, a 25m length is modeled with blind flanges at the end. For the entire model, external walls are adiabatic with no heat gain or loss. Hence conjugate heat transfer model is used for the CFD analysis. At start of valve opening (time=0), the temperature is initialized to ambient temperature, over the entire model and cold gas at temperatures well below -29C enters at the discharge flow rate.

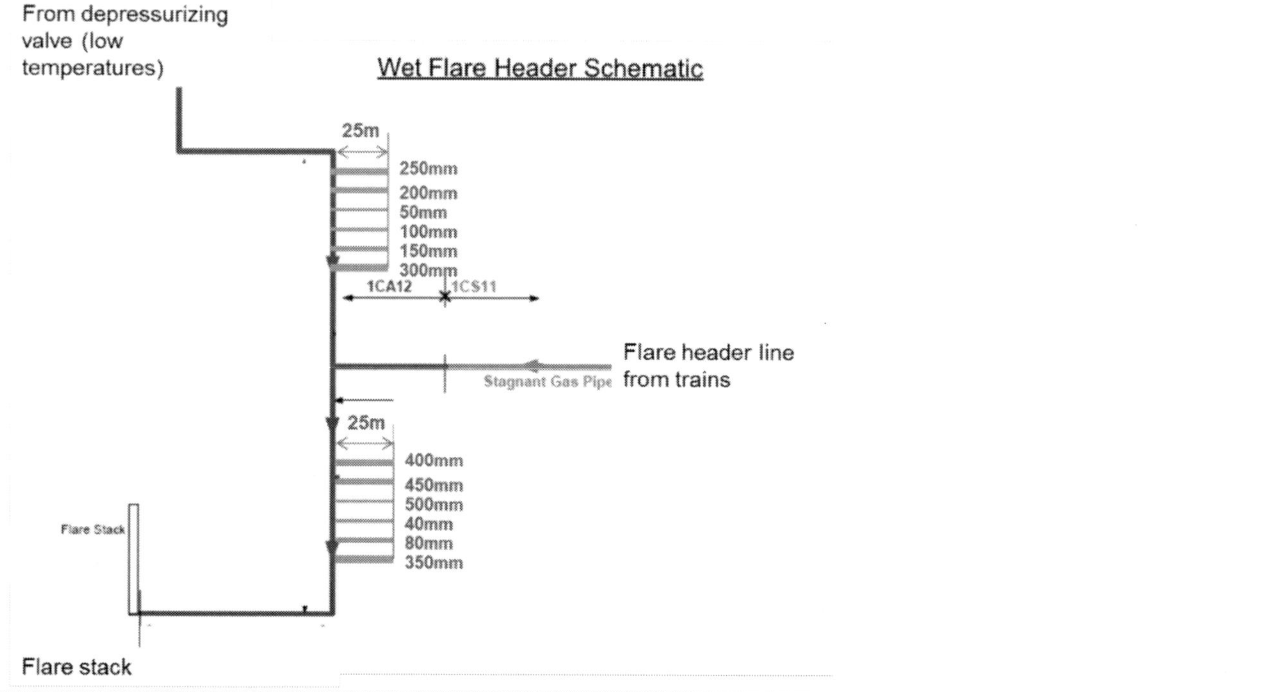

Figure 6: Schematic of flare header with various stagnant lines of varying diameters

The pressure reaches 2 bar in about a minute at which point the staging valve opens. Computation continues when the staging valve opens and the temperatures at various locations are monitored for 200 seconds allowing transients to settle down. Temperatures are then noted at steady state since only a slow monotonic decrease for some time is observed and then temperatures asymptote to a fixed value. Under steady state conditions, the conduction in metal is fast and inner and outer walls will show same temperatures. During the initial transient, CFD can capture conduction from inner to outer wall temperature. The radial thermal differential is critical for evaluating any potential crack formation. While the results hold good for steady state conditions, there are spikes during the packing and the initial part of the stage valve opening phase. Temperatures of about -35°C may be observed for short intermittent time along all diameter laterals but these are covered by the addendum to B31.3. Figure 9 shows the steady state temperatures for all the branches that were studied. From this figure it is possible to determine the length along the lateral after which it is suitable to transition to LTCS.

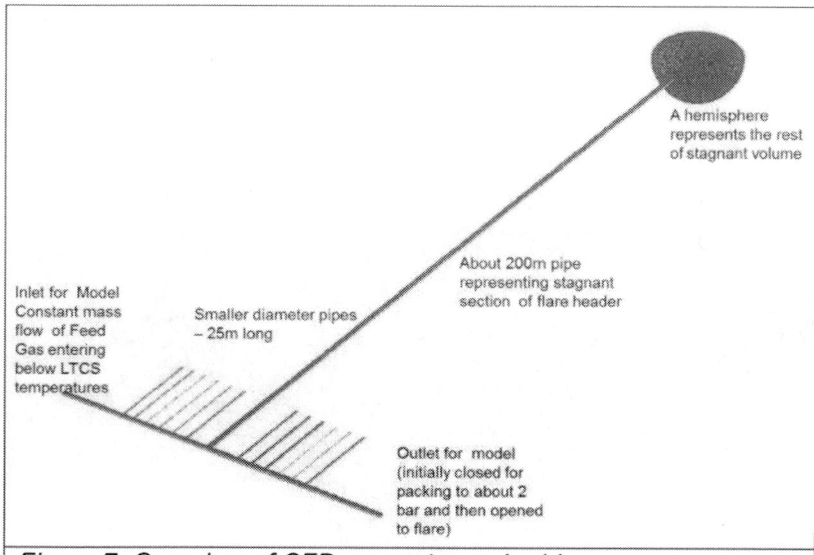

Figure 7: Overview of CFD geometry and grid

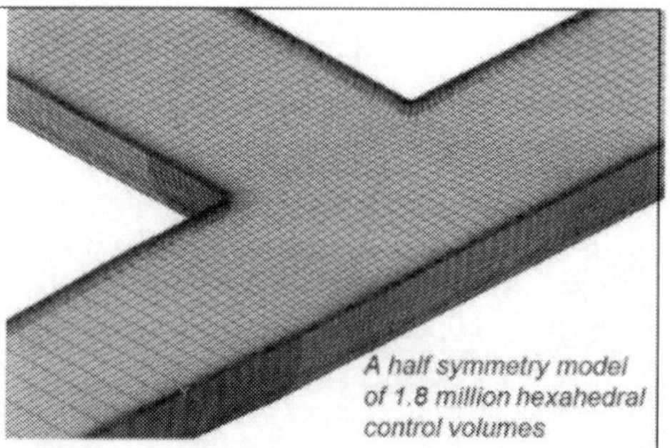

Figure 8: Boundary layer grid

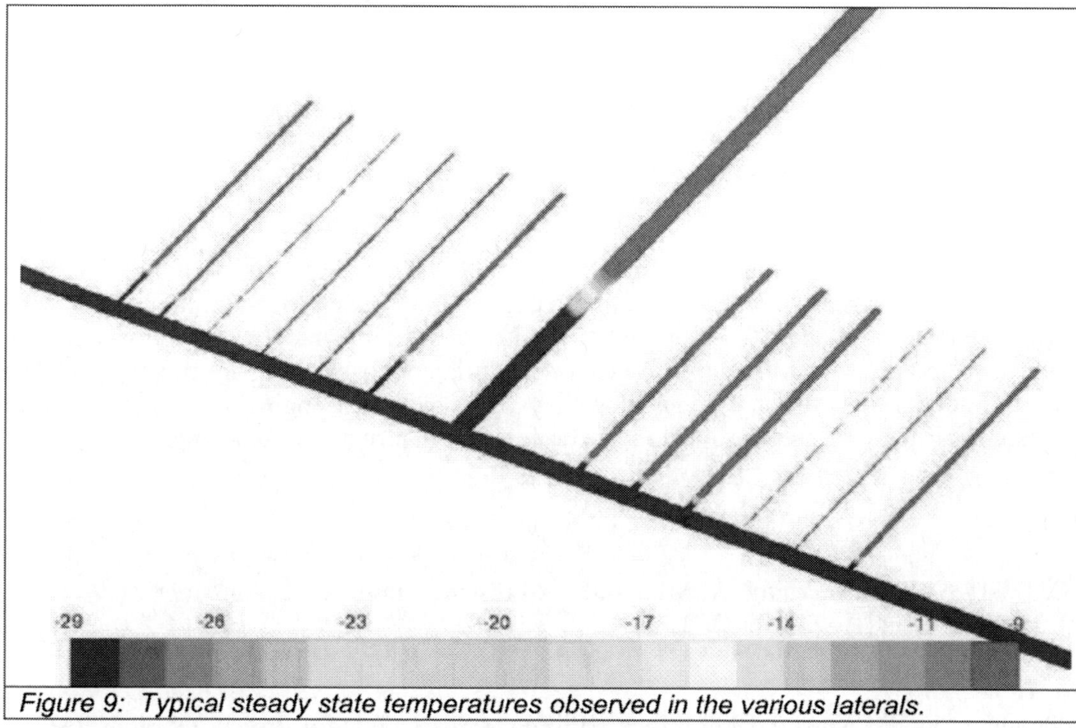

Figure 9: Typical steady state temperatures observed in the various laterals.

It takes about a minute to pack the volume to 2bar. The temperatures asymptote to a steady value after this point as shown in Figure 10. Figure 11 shows the temperature distribution along the inner walls of the pipe after 3 minutes and after 1 hour after the staging valve opens.

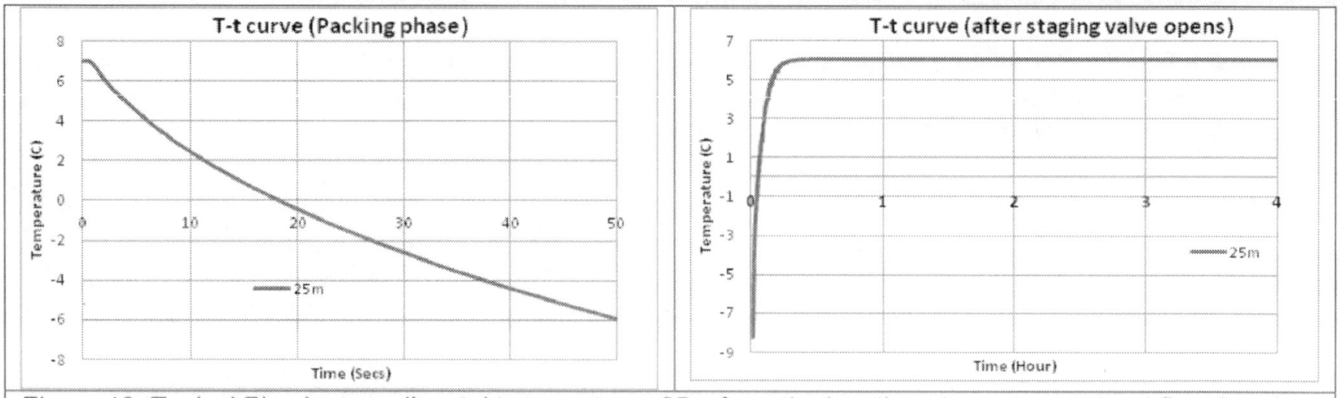

Figure 10: Typical Pipe inner wall metal temperatures 25m from the junction along stagnant wet flare header

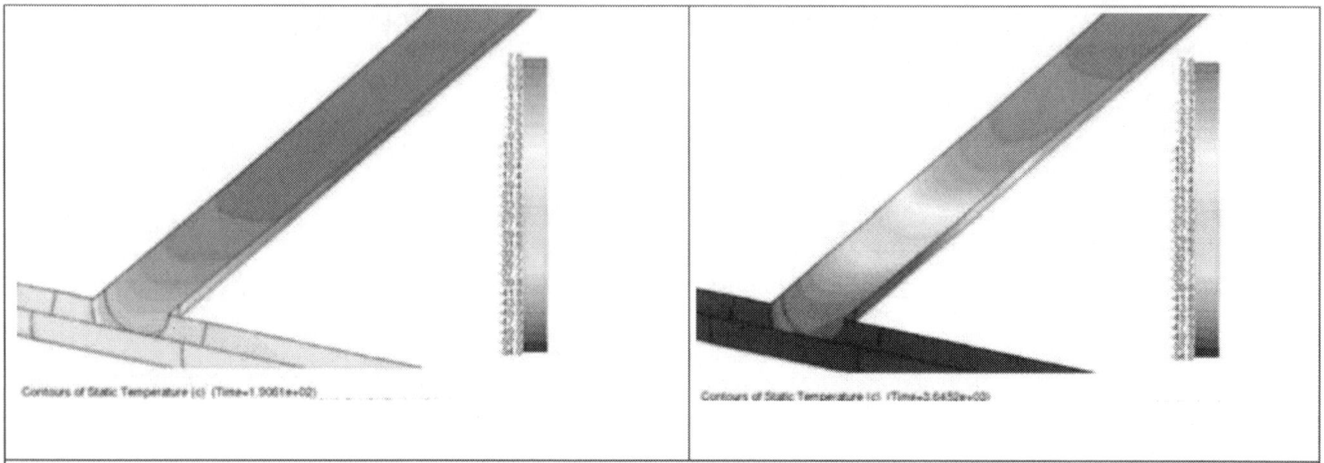

Figure 11: Temperature contours at 3 minutes and 1 hour after staging valve opens.

4. Conclusion

A combination of CFD and process simulations may be used for safe design and operation of flare headers and pressurization of lines downstream of main inlet feed isolation valves. Simulation should be deployed during the engineering stage of the project to assist in deciding material selection between stainless and carbon steel and the transition points to generate significant cost savings for the project. The versatility of analysis models and the unique insights they offer can help engineers design safe and sustainable processing facilities, as well as provide measurable value in terms of cost savings.

5. References

1. ASME B31.3-1999, Process Piping, ASME Code for Pressure Piping, B31, An American National Standard, Item (l) - Section 302.2.4.
2. ASME B31.3-1999, Process Piping, ASME Code for Pressure Piping, B31, an American National Standard, Table 323.2.2.
3. 2000 Addendum to ASME B31.3-1999, Process Piping, ASME Code for Pressure Piping, B31, American National Standard, Table 323.2.2 Notes and Figure 323.2.2B.
4. AS 2885.1-2007, pipelines - gas and liquid petroleum, Part 1: Design and construction, Section 4.9 – low temperature excursions.

444130 Computational Fluid Dynamics Simulation of Multi-Point Ground Flare Fields for Detailed Radiation and Plume Estimation

Abstract:

Multi-Point Ground Flare Fields are a critical component of Ethylene Production facilities, as they are the final emergency safety feature of the plant. Keeping personnel a safe distance from harmful radiation and deadly species concentrations is integral to the operation of these fields, and this is a primary concern in the design phase. Through Computational Fluid Dynamics, different flare tips can be compared in geometrically-detailed models that yield detailed estimates. Incident radiation and plume extents can be estimated for any operating case. Differing flare tip designs are validated against physical test data to ensure accuracy.

443979 GHS: OSHA's Final Deadline -- Last Call for Employer Compliance

Abstract:

OSHA's GHS transition period is set to end in 2016. By June 1 of this year, OSHA expects millions of HazCom-covered employers to be in full compliance with its fourth and final GHS deadline. This means having an updated HazCom Program in place that accounts for the GHS changes OSHA implemented when it adopted the United Nations' Globally Harmonized System in 2012.

This might sound straightforward, but adoption delays by chemical manufacturers and suppliers, as well as recent OSHA memorandums and selective enforcement rules, compromised earlier deadlines, and for many, created confusion in the chemical supply chain. Additionally, enforcement statistics have shown that just complying with the basic tenets of OSHA's HazCom Standard remains challenging for many employers.

In fact, the standard continues to rank #2 on OSHA's most frequently cited standard list. Compliance is important for avoiding fines and maintaining a positive corporate reputation, but it's probably most important for protecting employees from the increased injury and illness risks they face as a result of a compromised employer HazCom Program.

In this session, find out what it means to be in compliance with OSHA's June 1, 2016 employer deadline, what OSHA's general compliance expectations are, the common compliance challenges impacting EHS professionals and how to overcome them, and the actionable steps you can take to ensure compliance with OSHA's fourth and final GHS deadline and beyond.

444902 Rapid Assessment Tool for Process Safety Management (PSM) Systems

Abstract:

Senior management at facilities with significant process hazards have two key responsibilities related to process safety management (PSM). These responsibilities are:

(1) Ensure that the design of the management systems for process safety is adequate;

(2) Ensure that the management systems are implemented as designed.

As part of the second responsibility, full compliance audits are one method of evaluating PSM systems, but this type of audit can be resource-intensive. Besides, companies are reluctant to carry out compliance audits because of the cost and planning required for their execution. This presents a problem as areas of relatively high process risk may exist which are not effectively managed.

This presentation describes a simple assessment tool designed to rapidly and effectively identify areas where a process safety management system is doing well, and areas where opportunities for improvement in the system exist. Using the tool is not a substitute for a full assessment, but a way to ensure that process safety execution is going in the right direction, and addresses the areas with the most need for improvement first. The tool focuses on providing support to senior leadership.

The tool consists of 10 questions which are posed by an assessor who is an experienced process safety specialist, to be answered by high level site management and their representatives. The questions are based on generally accepted knowledge of the key drivers of a successful and self-sustaining PSM program. The answers provided to the questions, their analysis, discussion and the practical recommendations for improvement provided by the assessor will be useful to site management in designing and implementing a more effective PSM system.

442698 Safer and Smarter Sulfiding of Hydrotreating Catalysts

Abstract:

Catalyst sulfiding (aka "presulfiding") is a critical step to ensure that hydrotreating catalysts perform as designed after startup. In order to optimize the sulfiding process, Reactor Resources (RxR) has developed a new DMDS/TBPS Injection System that allows the refiner to continuously monitor sulfur injection rates, H2S levels, and Hydrogen Purity in real time during sulfiding operations. These mobile units, called SmartSkid™ Injection Systems, are equipped with wireless telemetry that is streamed in real-time via the cloud. With "live telemetry", engineers and operations personnel can monitor pump operations and recycle gas concetrations from any computer or smart phone.

This allows customers to constantly monitor parameters such as:

- Flow rate of the sulfur spiking agent
- Total quantity of chemical injected
- Pressure
- H2S Concentration
- Hydrogen Purity
- Fluid Temperature
- Valve Position
- Pump Speed

By utilizing an Online H2S Analyzer System to continuously measure the H2S concentration of the recycle gas, the hazardous practice of operations personnel taking draeger tube measurements while equipped with breathing air (SCBA) is eliminated. And with real time H2S data, sulfur injection rates can be controlled more precisely, avoiding over-injection of sulfur chemicals that can lead to compressor issues and the need to purge sour recycle gas. The online Hydrogen Purity Analyzer measures the hydrogen concentration of the treat gas in real time, which also helps to avoid compressor trips and excessive purging that can lead to excess sulfur emissions.

444112 Case Study of Accidental Chlorine Release in the Urban Area Using CFD and Acceleration of the Simulation

Abstract:

Computational Fluid Dynamics is widely used to simulate accidental hazardous chemical releases. In the urban area where dispersion obstacles are large and abundant, CFD would be the best choice to simulate the toxic chemical dispersion. In this paper, one industrial complex which is adjacent to residential area was chosen for the simulation and tested for 10 cases of the wind condition. One worst case was selected and solved for 1-hour after the chlorine release begins. And then for the vulnerable points in the simulation volume, toxic load and probability of death was calculated. After that, using Courant-Friedrich-Levy number, time stepping scheme of the implicit Euler method was modified to increase timte-linearly to make simulation faster but to keep robustness. With combination of linear CFL number and parallel solvers, CPU time of the CFD simulation became one-forth of the base case.

444116 Impacts of the New National Ozone Standard

Abstract:

Now that EPA has announced a new lower national ozone standard, there will be potential impacts to all sectors of the economy. This includes small industry, large industry, small businesses, large businesses, and potentially every individual consumer in the United States. This presentation will discuss the current state of the ozone air quality and the challenges to meet the new lower ozone standard. Challenges for industry located in ozone non-attainment areas include lower major source thresholds and higher NSR offset ratios for permitting. Challenges for state regulatory agencies include resources needed to address more non-attainment areas and addressing background ozone and international emissions contributions.

LUMMUS CONSULTANTS INTERNATIONAL

Towards a Level Playing Field

Permitting and Regulatory Compliance Around the World

Gordon Combs
Lummus Consultants International
(a CB&I Company)

A World of Solutions

LUMMUS CONSULTANTS INTERNATIONAL

Environmental Regulations in the USA

- In the beginning ...
  - ✓ 1948 – Federal Water Pollution Control Act ("FWPCA")
    - Established vague and often contradictory framework for national water pollution control policy
    - Established the Water Pollution Control Advisory Board
      - Surgeon General or his designee
      - Department of the Army representative
      - Federal Works Agency representative
      - Department of Agriculture representative
      - Six more Presidential appointees
        - Engineer with expertise in sewage and industrial waste disposal
        - Person with interest in wildlife conservation
        - Representative of municipal government
        - Representative of State government
        - Representative of an affected industry
        - One more person
    - Supported technological research into industrial waste treatment
    - Provided financial support to State and local entities for construction of treatment facilities

A World of Solutions April 2016 1

Environmental Regulations in the USA

- The Modern Era
  - ✓ 1970 – Environmental Protection Agency ("USEPA")
  - ✓ 1971 – Federal Clean Air Act ("CAA")
  - ✓ 1972 – Federal Clean Water Act ("CWA")
  - ✓ 1974 – Federal Safe Drinking Water Act ("SDWA")
    - Established underground injection control ("UIC") program
    - Does not apply to injection of brine or other fluids associated with the production of oil or natural gas
  - ✓ 1976 – Federal Resource Conservation and Recovery Act ("RCRA")
    - Established definitions of "solid waste" and "hazardous waste"
    - Completed the circle of regulations
      - Take it out of the air and put it into the water
      - Take it out of the water and put it into the ground
      - Take it out of the ground and put it into the air
- Regulatory system based on enforcement ("Command and Control")

International Regulations

- Europe and other "Developed" Countries
  - ✓ Comparable laws and regulations to the USA
  - ✓ Comparable promulgation schedule to the USA
  - ✓ Comparable enforcement policies to the USA
- Developing Countries
  - ✓ Comparable laws and regulations to the USA (today)
  - ✓ Promulgation schedules
    - Vary widely by country
    - Some nations still developing first regulations in the 1990s
    - Virtually all nations have comparable regulatory system today
  - ✓ Enforcement
    - Varies by country
    - More emphasis on working with industry to reach "win-win" solution

LUMMUS CONSULTANTS INTERNATIONAL

Project Finance

- Definition
  - ✓ Method of financing in which the lender looks primarily to the revenues generated by a single Project, both as the source of repayment and as security for the exposure, rather than to the parent company; i.e., "non-recourse" financing
  - ✓ Usually for large, complex and expensive Projects such as power plants, petroleum refineries, chemical processing plants, mines, oil and gas production, transportation, telecommunications and environmental treatment facilities
  - ✓ Financing of new Projects and refinancing of existing Projects
  - ✓ Borrower usually a Special Purpose Entity that is not permitted to perform any function other than developing, owning and operating the Project
  - ✓ Loan repayment depends primarily on the Project's cash flow and on the collateral value of the Project's assets
- Reserve-based financing in extractive sectors which is non-recourse and where the proceeds are used to develop one particular reserve (e.g., oil field) is considered to be a Project Finance transaction

A World of Solutions April 2016 4

LUMMUS CONSULTANTS INTERNATIONAL

Financial Institutions

- Multilateral Financial Institutions (Public Sector)
  - ✓ World Bank Group
    - International Bank for Reconstruction and Development ("IBRD") - government Projects
    - International Development Association ("IDA") – government Projects in poor countries
    - International Finance Corporation ("IFC") – loans for private Projects
    - Multilateral Investment Guarantee Agency ("MIGA") – political risk insurance
  - ✓ Regional Institutions
    - InterAmerican Development Bank ("IDB")
    - European Bank for Reconstruction and Development ("EBRD")
    - Asian Development Bank ("ADB")
  - ✓ Export Credit Associations ("ECAs")
    - United States Export-Import Bank ("US Ex-Im")
    - Export Development Canada ("EDC")
    - Japan Bank for International Cooperation ("JBIC")
- Development Banks (Private Sector)
    - Citigroup (USA)
    - Credit Agricole Corporate and Investment Bank (France)
    - Sumitomo Mitsui Banking Corporation (Japan)

A World of Solutions April 2016 5

World Bank Standards

- History
  - ✓ 1987 – Pollution Prevention and Abatement Handbook ("PPAH")
  - ✓ 1990s – Safeguard Policies
  - ✓ 2006 – Performance Standards
  - ✓ 2012 – Performance Standards (current revision)
- Today
  - ✓ Performance Standards – define borrower's responsibilities for managing the Project's environmental and social risks
  - ✓ Environmental, Health and Safety ("EHS") Guidelines – successor to PPAH
  - ✓ Common reference for Projects worldwide
  - ✓ All standards available via the Internet (for free!)
  - ✓ "Does the Project Comply with the World Bank Standards?"

IFC Safeguard Policies – 1990s

OP 4.01 Environmental Assessment

OP 4.04 Natural Habitats

OP 4.09 Pest Management

OP 4.36 Forestry

OP 4.37 Safety of Dams

OP 7.50 International Waterways

OP 4.20 Indigenous Peoples

OD 4.30 Involuntary Resettlement

OPN 11.03 Cultural Property

Child and Forced Labor Policy Statement

Lummus Consultants International

IFC Performance Standards - 2006

1. Social and Environmental Assessment and Management Systems
2. Labor and Working Conditions
3. Pollution Prevention and Abatement
4. Community Health, Safety and Security
5. Land Acquisition and Involuntary Resettlement
6. Biodiversity Conservation and Sustainable Natural Resource Management
7. Indigenous Peoples
8. Cultural Heritage

Lummus Consultants International

IFC Performance Standards – 2012

1. Assessment and Management of Environmental and Social Risks and Impacts
2. Labor and Working Conditions
3. Resource Efficiency and Pollution Prevention (reference the EHS Guidelines)
4. Community Health, Safety and Security
5. Land Acquisition and Involuntary Resettlement
6. Biodiversity Conservation and Sustainable Management of Living Natural Resources
7. Indigenous Peoples
8. Cultural Heritage

IFC EHS Guidelines

- Technical reference documents
  - ✓ Performance levels normally acceptable to IFC
  - ✓ Achievable in new facilities at reasonable cost by existing technology
  - ✓ Reference several globally accepted standards (e.g., WHO, USEPA, API, IMO)
- Project required to meet host country or IFC requirements
  - ✓ Whichever is more strict
  - ✓ Site-specific requirements if justified by Environmental Impact Assessment

General EHS Guidelines

- Cross-cutting EHS issues potentially applicable to all industry sectors
- Categories
  - ✓ Environmental
  - ✓ Occupational Health and Safety
  - ✓ Community Health and Safety
  - ✓ Construction and Decommissioning

LUMMUS CONSULTANTS INTERNATIONAL

Sector-Specific EHS Guidelines

- Industry-specific examples of Good International Industry Practice ("GIIP")
- Categories
  - ✓ Agribusiness / Food Production
  - ✓ Chemicals, including
    - Large Volume Petroleum-based Organic Chemicals Manufacturing
    - Natural Gas Processing
    - Petroleum Refining
  - ✓ Forestry
  - ✓ General Manufacturing
  - ✓ Infrastructure, including
    - Crude Oil and Petroleum Product Terminals
    - Ports, Harbors and Terminals
  - ✓ Mining
  - ✓ Oil and Gas
    - LNG liquefaction and regasification
    - Oil and gas exploration, development and production (onshore and offshore)
  - ✓ Power Generation, Transmission and Distribution

A World of **Solutions** April 2016 12

LUMMUS CONSULTANTS INTERNATIONAL

Equator Principles

- Purpose
  - ✓ Designed for use by private sector financial institutions – *particularly when one or more multi-lateral institutions is not involved*
  - ✓ Based on policies and guidelines of the World Bank Group
- History
  - ✓ 2003 – 10 banks from 7 countries created the Equator Principles
  - ✓ 2006 – First general revision
  - ✓ 2013 – Current revision (available via the Internet – for free!)
  - ✓ Today
    - 83 banks in 36 countries worldwide
    - 70 percent of international Project Finance debt in emerging markets

A World of **Solutions** April 2016 13

LUMMUS CONSULTANTS INTERNATIONAL

Equator Principles Categories and Requirements

1. Review and Categorization
   - ✓ Category A – potentially significant adverse environmental and social impacts
   - ✓ Category B – potentially limited adverse environmental and social impacts
   - ✓ Category C – minimal or no adverse environmental and social impacts
2. Environmental and Social Assessment
   - ✓ ESIA required for Category A and (some) Category B Projects
   - ✓ Alternatives analysis required if GHG emissions exceed 100,000 tonnes CO2-e per year
3. Applicable Environmental and Social Standards
   - ✓ Designated (developed) countries – host country standards
   - ✓ Non-designated countries – IFC Performance Standards and EHS Guidelines
4. Environmental and Social Management System and Action Plan
5. Stakeholder Engagement
6. Grievance Mechanism
7. Independent Review – pre-financial close
8. Covenants – Borrower agrees to comply (in writing)
9. Independent Monitoring and Reporting – construction and operations
10. Reporting and Transparency

A World of **Solutions** April 2016 14

LUMMUS CONSULTANTS INTERNATIONAL

Conclusions

- Today, industry cannot "export pollution" by building facilities in developing countries with lax environmental and social requirements
- Some international requirements (e.g., offsite noise control) are stricter for Projects in developing countries than in the USA
- Mitigation of potential social impacts just as important (if not more important) as mitigation of potential environmental impacts
- Financial institutions are a primary driving force behind international environmental and social progress associated with public and private industrial development
- Golden Rule of Business: "Whoever has the gold makes the rules!"

A World of **Solutions** 15

Lummus Consultants
International

CB&I

Lummus Consultants
International

A World of **Solutions**

BRACEWELL

AIChE Ethylene Producers' Conference
Ethylene Plant Environmental Panel

Proposed Amendments to EPA's Risk Management Program Regulations

April 12, 2016

Matt Paulson
Bracewell LLP

BRACEWELL

Overview

- Background on EPA's Risk Management Program
- Overview of Existing RMP Regulations
- Review of Proposed Rule
  - Prevention Program Requirements
  - Emergency Response Preparedness Requirements
  - Information Availability Requirements
- Specific Requests for Comment

2

BRACEWELL

Background on Risk Management Program

- 1984, Union Carbide, Bhopal India
- 1985, Waxman (D. CA) introduces first proposed legislation
- 1986 EPCRA
- 1990 CAA Amendments

3

BRACEWELL

Overview of Existing RMP Regulations

- Applicability
  - More than a threshold quantity
  - Of a regulated substance
  - In a process
  - At a stationary source
- Three Main Components
  - Hazard assessment
  - Prevention program
  - Emergency response program

4

BRACEWELL

Overview of Existing RMP Regulations (cont.)

- Three-Tiered Program
  - **Program 1**: No affect to public of worst case release; no accidents with offsite consequence in five years; limited to hazard assessment, coordination, and plan submission
  - **Program 2**: Not eligible for Program 1/not subject to Program 3; streamlined prevention program requirements; additional hazard assessment, management and emergency response requirements
  - **Program 3**: Not eligible for Program 1 and either subject to OSHA PSM standard or listed industry sector; accident prevention requirements (14 PSM elements); Program 2 requirements

5

BRACEWELL

Background on Proposed Rule

- Why is EPA revising the RMP regulations now?
  - 2013 West Fertilizer incident
- The following incidents are also highlighted by EPA in the preamble:
  - BP Texas City
  - Tesoro Anacortes
  - Chevron Richmond
  - Williams Geismar

6

BRACEWELL

Background on Proposed Rule (cont.)

- August 2013 – Executive Order 13650
- December 2013 – OSHA PSM RFI
- May 2014 – Working Group Report
- July 2014 – EPA RMP RFI

7

BRACEWELL

Background on Proposed Rule (cont.)

- July 2014 – EPA RMP RFI (cont.)
  - Areas addressed in RFI but not covered by proposed rule:
    - Regulated Substances List
    - RAGAGEP
    - Safety Critical Equipment
    - Safety Case
    - Facility Siting
    - Automatic Detection and Monitoring
    - Offsite Consequence Data
    - Emergency Shutdown

8

BRACEWELL

Background on Proposed Rule (cont.)

- Dec., 21, 2015 – EPA proposed rule sent to OMB
- February 25, 2016 – Pre-publication version of proposal released
- *Federal Register* publication – March 14, 2016
- Comment deadline – May 13, 2016
  - Unusual, shorter "soft" deadline for information availability-related comments – April 13, 2016
- Final rule issued – Fall 2016?

9

BRACEWELL

Overview of Proposed Rule

- Prevention Program Requirements
  - Incident Investigation
  - Third Party Compliance Audits
  - Safer Technology and Alternatives Analysis
- Emergency Response Preparedness Requirements
  - Coordination with Local Responders
  - Facility Exercises
- Information Availability Requirements
  - LEPC Disclosure Requirements
  - Public Availability of Information
- Specific Requests for Comment

10

BRACEWELL

Prevention Program Requirements

- Incident Investigation
  - Revisions to Definition of "Catastrophic Release"
    - "The owner or operator shall investigate each incident that ~~which~~ (1) resulted in~~, or could reasonably have resulted in~~ a catastrophic release ~~of a regulated substance~~ . . . or (2) Could reasonably have resulted in a catastrophic release (i.e., was a near miss)."
    - "Catastrophic release means a major uncontrolled emission, fire, or explosion, involving one or more regulated substances that ~~presents imminent and substantial endangerment to public health and the environment~~ results in deaths, injuries, or significant damage on-site, or known offsite deaths, injuries, evacuations, sheltering in place, property damage, or environmental damage.

11

BRACEWELL

Prevention Program Requirements (cont.)

- Incident Investigation (cont.)
  - New definition – "root cause"
    - The report shall include . . . "factors that contributed to the incident including the initiating event, direct and indirect contributing factors, and root causes."
    - "Root cause means a fundamental, underlying, system-related reason why an incident occurred that identifies correctable failure(s) in management systems."
  - Use of undefined term – "near miss"
    - Preamble discussion: "any unanticipated release of a regulated substance, and some process upsets could potentially lead to a catastrophic release."

12

BRACEWELL

Prevention Program Requirements (cont.)

- Third Party Compliance Audits
  - When is it required?
    - RMP-reportable accident, or
    - Regulatory authority requires one based on any noncompliance
  - Auditor Qualifications
    - RMP program knowledge, certification, training and PE
    - Source type and process unit experience
    - Experience with applicable RAGAGEP
    - Receive no financial benefit tied to outcome
    - No consulting services for owner/operator in past three years or for three years post-audit

13

BRACEWELL

Prevention Program Requirements (cont.)

- Third Party Compliance Audits (cont.)
  - Audit Report
    - Must provide to EPA
    - Include drafts
    - No legal privileges – becomes publicly accessible
  - Audit Findings
    - 90 days after completion of audit owner/operator must submit response to each finding with a schedule for "promptly" addressing deficiencies
  - Submission to Audit Committee of Board of Directors

14

BRACEWELL

Prevention Program Requirements (cont.)

- Safer Technology and Alternatives Analysis
  - Applies to Program 3 facilities in three industries:
    - paper manufacturing
    - coal and petroleum products manufacturing
    - chemical manufacturing
  - Facilities are required to evaluate the "feasibility" of safer technology and alternatives when conducting PHAs

15

BRACEWELL

Prevention Program Requirements (cont.)

- Safer Technology and Alternatives Analysis (cont.)
  - PHA must include analysis of potential safer technology and alternatives in the following order of preference: IST or ISD, passive measures, active measures, procedural measures
    - IST/ISD: measures that would replace or reduce the use of regulated substances or make operating conditions less hazardous or complex
    - Passive measures: dikes, berms, blast walls
    - Active measures: alarms, safety instrumented systems
    - Procedural measures: policies, operating procedures, training

16

BRACEWELL

Prevention Program Requirements (cont.)

- Safer Technology and Alternatives Analysis (cont.)
  - Definition of "feasibility": "capable of being successfully accomplished within a reasonable time, accounting for economic, environmental, legal, social and technological factors."
  - EPA rejected this same concept in 1996 rulemaking – concluded that the existing PHA requirement was sufficient.
  - What has changed?
    - Safer technologies are "becoming more widely available and being adopted by some companies"

17

BRACEWELL

Prevention Program Requirements (cont.)

- Prevention Program Takeaways:
  - Potential for significant increase in number of formal investigations
  - Mandates identification and correction of management system failure even where no such failures exist
  - Releases with only on-site impacts can trigger formal investigation
  - Third party audits could be required in response to any violation
  - Disclosure of audits could fundamentally undermine audit process
  - Rigid auditor qualifications could exclude many auditors
  - STAA would require burdensome justifications for technology and design choices

18

BRACEWELL

Emergency Response Preparedness

- Coordination with Local Responders
  - Documentation of coordination efforts:
    - Names and contact information of individuals involved
    - Dates of coordination
    - Conclusions
    - Next steps

19

BRACEWELL

Emergency Response Preparedness (cont.)

- Facility Exercises
  - Annual tabletop emergency response exercises
  - Field exercise every five years
  - Annual notification exercises

20

BRACEWELL

Emergency Response Preparedness (cont.)

- Emergency Response Takeaways:
  - Potential to overwhelm local responders
  - Owner/operator would bear enforcement risk – "ultimate burden rests with the owner/operator"
  - Security concerns

21

BRACEWELL

Information Availability Requirements

- LEPC Disclosure Requirements
  - Regulated substances on site
  - Compliance audit summary reports
  - Summaries of incident investigation reports
  - Summary if IST implemented or planned to be implemented
  - Reports on emergency response exercises/drills

22

BRACEWELL

Information Availability Requirements (cont.)

- Public Availability of Information
  - Names of hazardous chemicals
  - MSDSs for all regulated substances at facility
  - Five-year accident history
  - Summary of emergency response program
  - Summary of emergency response exercises
  - LEPC contact information

23

BRACEWELL

Information Availability Requirements (cont.)

- Information Availability Takeaways:
  - Public already has access to much of this information
  - Public release of sensitive security information
  - Unclear how much detail is required in "summaries"

24

BRACEWELL

Specific Requests for Comment

- Increase scope of RCA
- Require incident investigation report completed before restart
- Expand third party audit trigger to include conditions "suggesting a concern"
- Require third party auditor to submit draft report to regulatory authority before provided to owner/operator
- Require implementation of STAA, rather than feasibility analysis
- Create "score card" for third party auditors to grade facilities
- New facility siting requirements (extensive discussion)
- Emergency shutdown (extensive discussion)

25

BRACEWELL

Specific Requests for Comment (cont.)

- Request for Comment Takeaways:
  - Could provide EPA with argument against claim that adding new requirements to final rule not in proposal violates scope of proposal/reasonable outgrowth limitations
  - Facility siting and emergency response most likely to result in new requirements in final rule that were not in proposal

26

BRACEWELL

Matt Paulson
Bracewell LLP

111 Congress Avenue, Suite 2300
Austin, Texas 78701-4061
512.494.3659

711 Louisiana Street, Suite 2300
Houston, Texas 77002-2770
713.221.4959

matthew.paulson@bracewelllaw.com

Texas
New York
Washington, D.C.
Connecticut
Seattle
Dubai
London

bracewelllaw.com

27

Application of Dynamic Simulation in Improving Plant Efficiency

By

Tim Song, Maryn Pols, Prakash Ramdoss

CB&I
2 Riverway, Suite 1300
Houston, TX 77056
Tel: +1 713 375 8711

Prakash.ramdoss@cbi.com

Abstract

In ethylene plants, considerate amount of CO2 and NOx emissions come from gas turbine exhaust gas and furnace flue gas. The hot gas turbine exhaust gas may be reused as oxygen source in the thermal cracking furnaces in place of cold ambient air, thus increasing the overall thermal efficiency, potentially reducing CO2 and NOx emissions from gas turbine combustion chamber and furnace combustion chamber combined. However the integration between gas turbine operation and the furnace operations raises operating flexibility and plant safety concerns, as the combined process needs to be able to operate in both ambient air mode and turbine exhaust gas mode and be able to transit from one mode to another in a safe and robust manner.

This work is part of a revamping project at an ethylene plant, where a new gas turbine is to be installed and integrated with the existing furnaces to improve plant efficiency. The hot gas turbine exhaust gas is routed to the furnaces, instead of the cold ambient air there by reducing fuel need for the furnaces. The work demonstrates that the dynamic process simulation models can be utilized to study the transition between modes of operation, revealing critical process insights. The transient analysis can also be used to determine key operating parameters such as damper opening and closing time parameters to ensure safe and smooth transitions and operations. Consequently, the analysis assures the plant operators with high confidence the robustness of integration between the gas turbine and furnace processes which increases the plant fuel efficiency thereby reducing emissions.

Introduction

The flue gas from the new gas turbine is reused as oxygen source for the thermal cracking furnaces, therefore reducing CO2 and NOx emissions from gas turbine exhaust gas and furnace flue gas combined.

A transient analysis for the flue gas system is required to ensure that sufficient combustion air flow to the cracking heaters is maintained at all times during the system transient conditions to prevent sub-stoichiometric combustion and positive heater firebox pressure leading to personnel safety concerns.

The purpose of the computer modeling is to dynamically represent the combustion air supply system and induced draft flue gas system to allow simulation of the transient response of these systems during upset conditions.

The computer model determines the pressure, temperature, mass flow rate and composition at critical points in the system as a function of time. The interaction of the combustion air, the louvers and dampers and the controllers is very important to an accurate calculation of the dynamic response of the integrated combustion air system. The time delays associated with the transient phenomena in the ducting, the sensor response and the valve motions are modeled to determine if the control systems can respond fast enough to prevent an unstable or dangerous condition. The residual heat in the refractory-lined duct system is taken into account for an accurate representation of the flows. Boundary conditions for the computer analysis are provided by models representing the fans, gas turbine, louvers and dampers, controllers and cracking heaters.

The transients of the upset scenarios described below are analyzed for the new gas turbine duct system.

1. The new gas turbine trip.
2. Trip of trim fan associated with the new gas turbine.
3. Trip of the induced draft fan on a thermal cracking furnace.

The primary purpose of the transient analysis is to ensure that the post-trip interlock logic and control strategy:

- Provides sufficient combustion air flow to the cracking heaters at all times during the system transient conditions avoiding shutting down heaters.
- Prevents sub-stoichiometric combustion.
- Prevents a severe pressure transient which may temporarily lead to positive heater pressures causing concerns regarding personnel safety.
- Avoids potential damage to the combustion air and gas turbine exhaust ducts due to excessive draft pressure.
- Prevents the trip of the GTG as a result of a single heater upset.
- Prevents potentially destabilizing pressure oscillations in the system.

The transients were performed using a computer dynamic simulation model of the new gas turbine combustion air system and induced draft flue gas system which includes all principal components (gas turbine, trim fan, combustion air supply ducting and stack, heater radiant sections, heater convection sections, ID fans and FD fans), control dampers and associated control systems.

The results have been utilized to implement appropriate post-trip interlock logic and control strategies for the revamping project.

System Description

The existing heater island consists of a number of thermal cracking furnaces, with half of them are already integrated with an existing gas turbine, supplemented with ambient air. In order to improve overall plant energy efficiency, a new gas turbine exhaust gas will be reused for the rest of thermal cracking heaters currently using ambient air for combustion.

The existing turbine exhaust gas duct system will be modified and expanded to accommodate the integration of the new machine with the thermal cracking heaters currently operating on 100% ambient air. An overflow connection between the new duct system and the existing duct system will be installed to maximize the turbine exhaust gas utilization in case a heater on the new duct system is shut down. The new duct system will have one operating trim fan without spare.

All thermal cracking furnaces have induced draft fans at the furnace top, while some of them have forced draft fans at air inlet and some do not. Upon a new gas turbine trip, some thermal cracking heaters will trip to forced draft operation while others will trip to natural draft operation. The duct branch to each of these heaters will be equipped with an ambient air intake system.

The ID fans control the pressure at the firebox arch. The furnaces that do not have forced draft fans will each be equipped with a dedicated air intake duct with control damper, which will allow ambient air operation using the ID fan as driving force to pull in combustion air when the gas turbine is down.

The furnaces require additional oxygen for stoichiometric combustion and the trim fan by side of the turbine exhaust duct provides the additional oxygen by mixing ambient air with turbine exhaust gas in the mixer. The turbine exhaust stack provides relief during upset events leading to high pressure in the duct. The main duct is pressure controlled by manipulating trim fan inlet damper, while the trim fan operates at constant speed. All furnaces are lined up along the main exhaust duct. The cross flow duct with a cross damper connects the new duct system to the existing GTG system. Normally the cross flow duct is closed. When one or more furnaces in the new duct system are down, the surplus exhaust gas will be routed to the existing duct for better plant efficiency. The furnaces' FD fans and FD fan dampers during normal turbine exhaust gas operation are closed and the fans are not running. Upon a trip of the new turbine, the furnaces will trip to forced draft operation if there is a forced draft fan associated with it, otherwise natural draft ambient air operation.

Within each heater, the simulation model takes into account the buoyancy effect, burner combustion air side pressure loss, and the convection section hydraulics loss at flue gas side in addition to the induced draft fan and the fan damper characteristics.

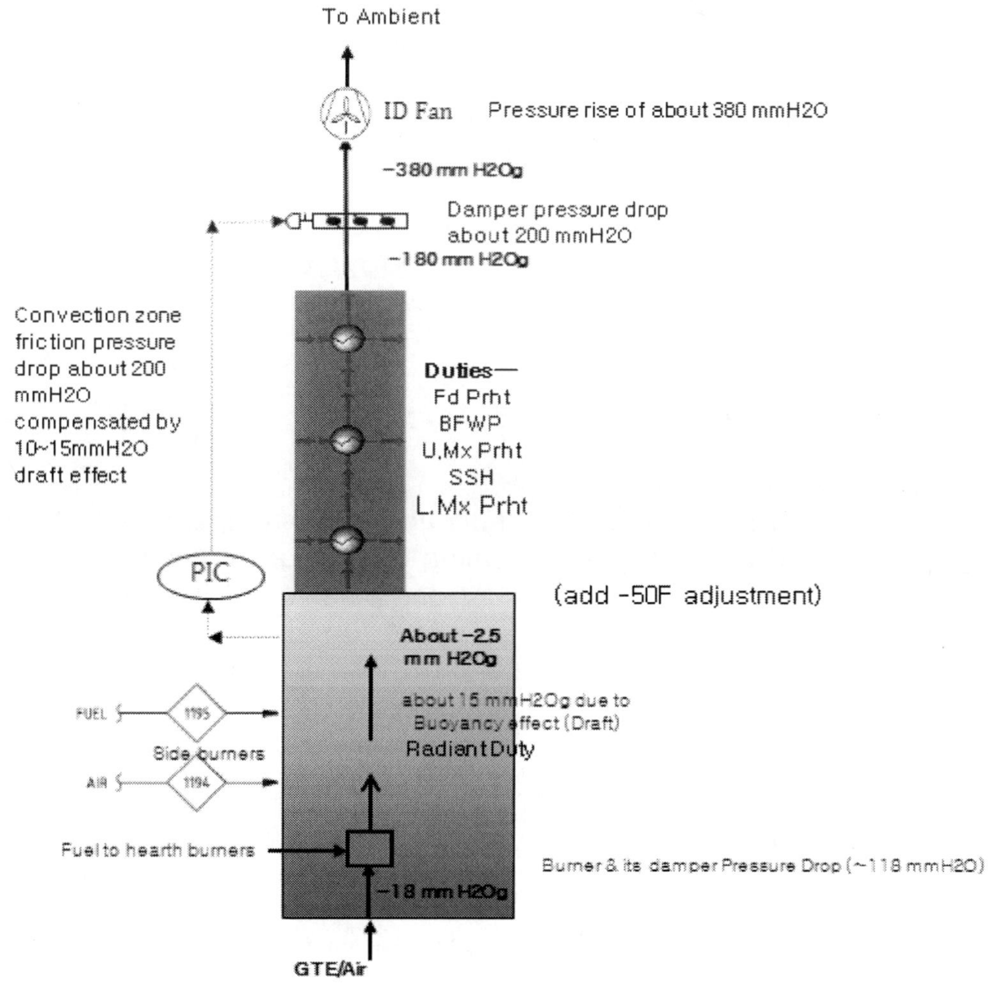

When gas turbine is tripped, both of the flow and temperature decay over time. In order to predict the pressure loss and temperature effect during transition from turbine gas operation to ambient operation and vise versa, duct sizes and duct metal wall thickness, as well as insulation layer are considered in simulation model.

All fans and dampers are modeled with performance curves defined by flow vs pressure head/loss.

Within each furnace, the process side duty demand is correlated to flue gas side by cross limiting control mechanism, such that the coil outlet temperature is properly controlled. The process side feed rate to each furnace is assumed to remain constant during the transients.

On the flue gas side of each furnace, the arch is pressure controlled by the overhead ID fan inlet damper while the ID fans operate at constant speed.

Buoyancy is rigorously considered in the firebox as well as in the flue gas stack according to past engineering practices.

Combustion is taking place in the firebox. The combustion is rigorously calculated thermodynamically, that is, the combustion reaction ensures atomic balance. If there is shortage of oxygen, carbon monoxide will be generated.

The furnace convection section is rigorously modeled at the flue gas side with actual flue gas path volume and measured heat transfer coefficients.

Process side modeling is mostly simplified with duties proportional to the feed rate and inlet temperature. The purpose of modeling feed temperature effect is such that the coil outlet temperature can eventually be adequately predicted and used as input signal to cascade control the firing via cross limiting control mechanism.

Results and Discussion

Gas Turbine Trip

The gas turbine trip is imposed 10 seconds into the simulation. At 10 seconds, the turbine is tripped and the corresponding events are carried out by the simulation model, primarily closing flue gas duct dampers to each furnace while jump open induced draft fan damper opening of each furnace momentarily. The ambient air entry to the furnaces without forced draft fans will be active when duct pressure becomes negative.

During gas turbine trip the concern is the shortage of flue gas/air, therefore oxygen rate, going into each furnace. The simulation predicts the flow rates, as well as oxygen rates, going into a typical furnace are as follows -

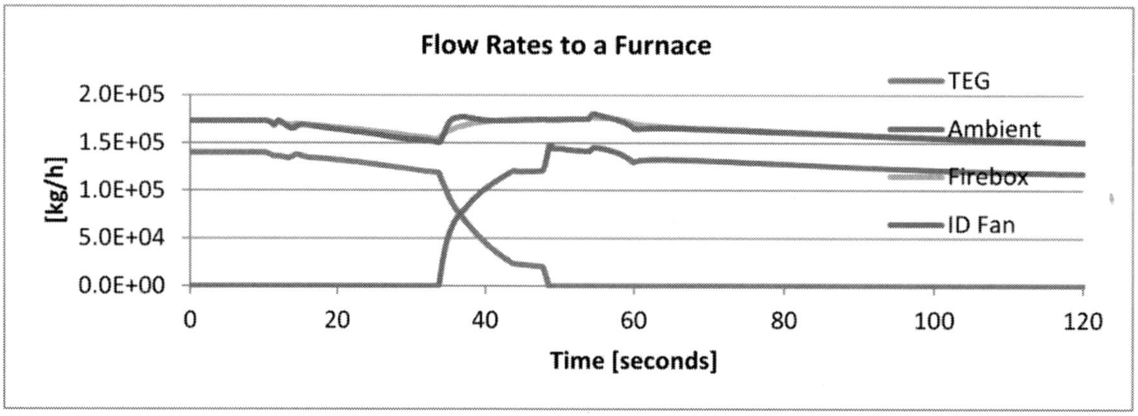

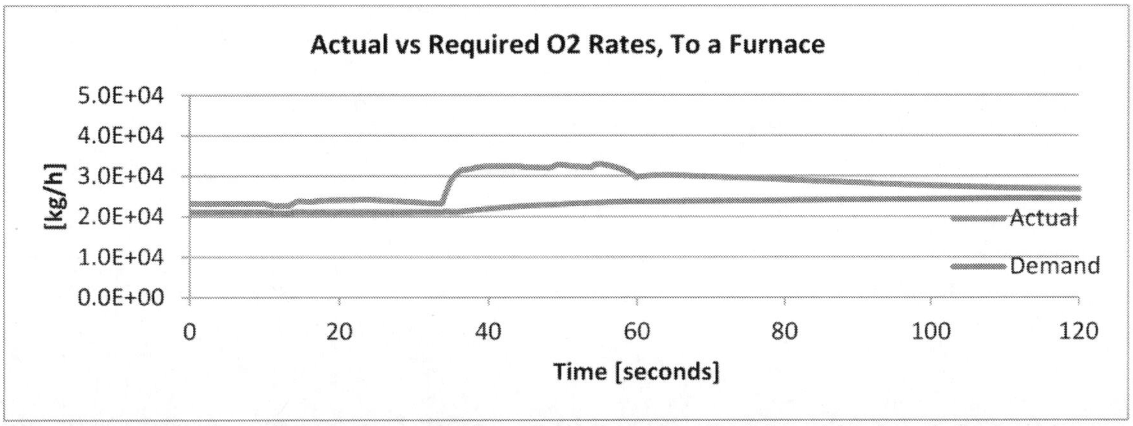

As observed, the simulation verifies that the gas turbine tripping interlocking actions are adequate to ensure sufficient air/oxygen supply to furnaces. As such, sub stoichiometric combustion has been avoided, while production rates remain unchanged.

Trim Fan Trip

The trim fan trip is imposed 10 seconds into the simulation. The sequence of events is implemented according to definition, primarily starting up the forced draft fans for those furnaces with one while leaving the rest of the furnaces intact.

The concern is the potential shortage of oxygen in the furnaces' firebox and positive pressure in the furnaces with forced draft fans.

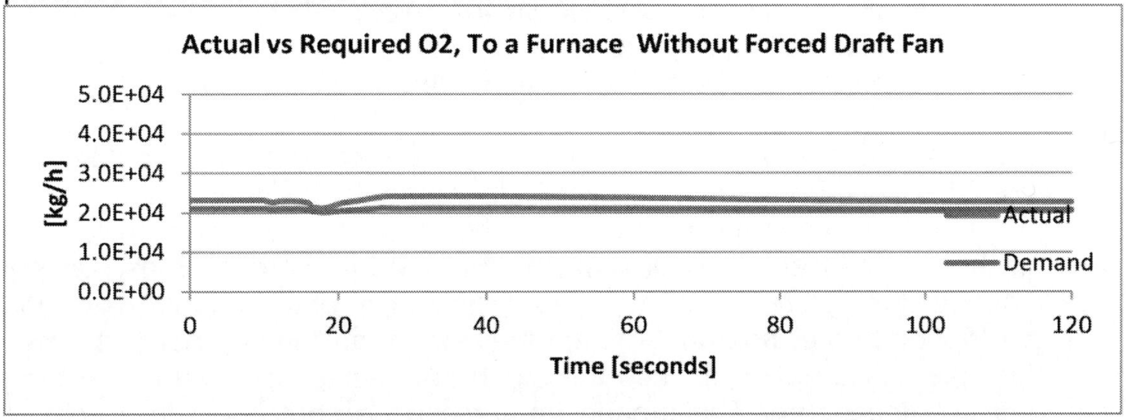

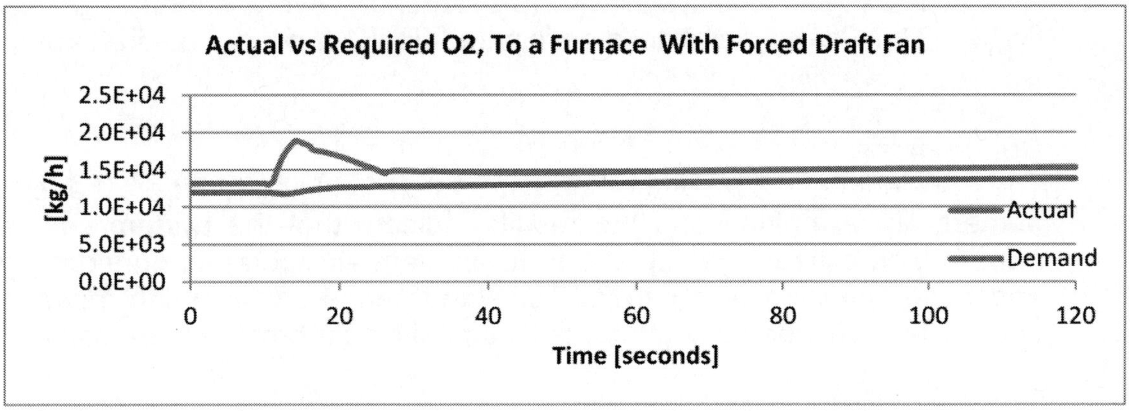

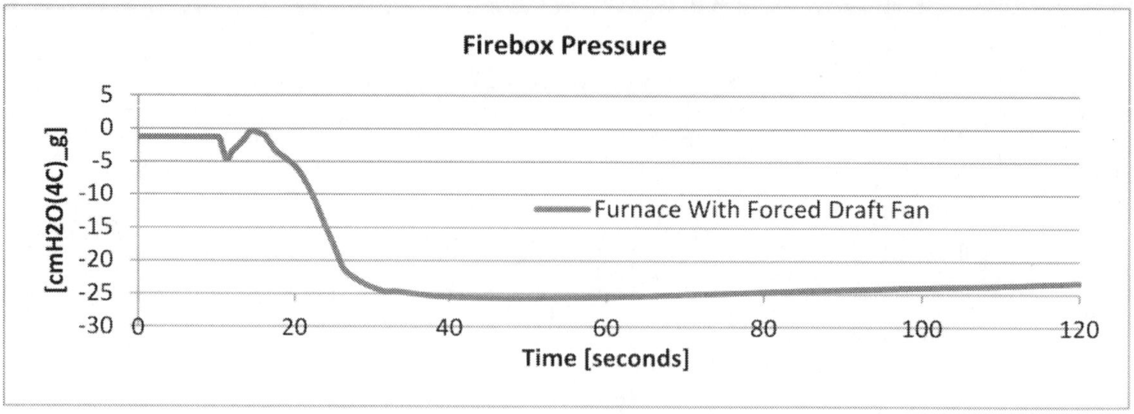

As seen, for a furnace without a forced draft fan, the oxygen supply reaches minimal level 5 seconds into the trip, yet still remains over the oxygen demand at the time. The oxygen supply is ample for furnaces with forced draft fans and the fire box pressure is also below ambient, with the highest pressure at about 5 seconds into the trip.

Induced Fan Trip

The induce fan trip is imposed 10 seconds into the simulation. Some details actions of the trip are listed below:

1. The induced draft fan is tripped and the fan is spinning down in about 3 minutes.
2. Trip close process side feed valve.
3. Trip close all fuel valves.
4. Opening up process side dilution stream valve by 25%.

Since the induced fan takes quite some time to come down, the flue gas damper will stop the flow from furnace inlet side much faster than the fan spinning down process. The net result is that the firebox pressure is predicted to go sub-ambient for as low as -150 mmH2O(g) for the furnace that is tripped, as consequence of its induced fan trip. However the abrupt stop of turbine exhaust gas consumption by the tripped heater does not disrupt the rest of the running heaters. The reduction in flue gas consumption causes the stack damper to open and vent the surplus flue gas to atmosphere. The tripped heater can be shut down in a controlled manner without leading to positive firebox pressure or severe shortage of oxygen in the firebox. Thus, safe operation is anticipated during the induced draft fan trip.

Conclusions

The transient behaviors of the flue gas system bridging turbine exhaust and furnace burners are examined and the results indicate that the system can safely transit from turbine exhaust gas operation to ambient air operation during gas turbine trip, under the condition of a carefully selected series of mitigation measures. That is, both sub-stoichiometric combustion and positive firebox pressure can be avoided.

The results also indicate that the system can safely transit from the induced draft fan trip of a heater, and trim fan trip as long as proper mitigation measures are implemented.

Highlights of the key mitigation measures during gas turbine trip include:

1. For each furnace, switch arch pressure controller and TEG/air intake flow controller to manual mode and open ID fan dampers and TEG dampers on heaters without FD fans to a certain degree while holding TEG dampers opening for heaters with FD fan.

2. When FD fan speed is up, open FD fan damper to a certain degree while closing the corresponding TEG damper, making sure the net flow to the heater remain largely unchanged.

3. When the main duct pressure drops below ambient, open stack damper fully; shutdown trim fan and close its damper; open up ambient air intake dampers to a certain degree. At same time, close TEG dampers such that the net air flow to heaters largely remains constant.

4. Optionally adjust air dampers opening to manage air flows into heaters.

5. Eventually when the process in each heater is sufficiently stable, switch the arch pressure controller back to auto, restore the air damper controls to auto and conclude the transition.

Highlights of trim fan trip are as follows:

This is an event with lesser impacts than the gas turbine trip event. The key measure is to slowly ramp close the TEG damper in about 20 seconds while allowing the combustion air flow controllers to automatically open the ambient damper after a short delay to allow the FD fan discharge pressure to equal with the duct pressure. As a result of operating on FD fans, the surplus TEG will eventually be vented to atmosphere through the stack damper. The firebox pressures can be adequately maintained below ambient while no oxygen shortages are seen.

Highlights of ID fan trip are as follows:

When the fan of a furnace trips, the TEG damper is tripped closed, stopping the combustion air flow into the tripped heater. The remaining running heaters can effectively reject the surplus TEG causing duct pressure to rise to the point that the stack damper pressure controller set point is reached and venting through stack takes place. Positive pressure is not observed in the tripped heater firebox due to residual capacity of the tripped fan during the long period of fan rotor decay time while the faster TEG damper stroke time stops the TEG flow to the heater in just a few seconds, which causes firebox pressure to go down to as low as -150 mmH2O(g).

<u>**References**</u>

1. Transient Analysis Duty Specification
2. Basis of Design
3. Material Balance
4. Process Description
5. Equipment List
6. Hydraulics Summary
7. Process Control Description
8. PFD and P&ID
9. Interlock Summary
10. Convection section rating tool reports
11. Damper characteristic definition
12. Plot plan
13. Duct Pressure Loss Calculations
14. Reference projects done in the past.
15. Heater burner data sheets
16. Heater data sheets
17. KOM
18. TRIM FAN DAMPER Perf. Curve
19. Pressure Drop Calculation
20. Damper with Actuator Data Sheet
21. Technical Data Table for Process Damper
22. FDF datasheet & curve
23. MOM Study Review Meeting
24. DCS Data for Analysis
25. Actual ID fan damper opening

444054 Thermodynamic Modeling of CO_2 Solubility in Aqueous Solutions of NaCl, KCl, $CaCl_2$, $MgCl_2$ and Their Mixed Salts Using ENRTL Model

Abstract:

The influence of electrolytes on phase equilibria of CO2 is important to various chemical, petroleum, and environmental applications including CO2 capture and sequestration. A thermodynamic model based on the electrolyte NRTL activity coefficient equation and PC-SAFT equation-of-state is developed for CO2 solubility in aqueous NaCl, KCl, CaCl2, and MgCl2solutions at temperatures from 273 to 473 K, pressures up to 150 MPa, and salt concentrations up to saturation. Effects of different electrolyte species on CO2 solubility in aqueous solutions have been investigated. The eNRTL binary interaction parameters for the CO2–ion pair are regressed against the experimental CO2 solubility data for aqueous single electrolyte solutions. Additionally, the solubility of CO2 in aqueous solutions of mixed electrolytes including NaCl + KCl, NaCl + CaCl2 , KCl + CaCl2), NaCl+MgCl2, KCl+MgCl2 and CaCl2+MgCl2 (each at weight ratio = 1:1) are predicted with the eNRTL binary parameters. the prediction results favorably agree with the experimental values.

443899 Examination of the Influences on the Absorption of SO_2 into Seawater in a Packed Mini Column and a Pilot Column

Abstract:

In fossil-fueled power plants, which are located at the coast, seawater can be used for flue-gas desulfurization as already successfully accomplished throughout the last years.

Due to the stricter regulations, in the last years ships are often furnished with flue-gas desulfurization plants which can also use seawater as absorbent. As there is only very limited space for flue-gas scrubbers at the ships, it is important that they are working efficient. Therefore, the use of random and structured packings is promising as they provide a huge surface area for the interaction between the two phases.

Furthermore, the composition of the seawater is also an important factor, which has to be considered as it varies with the geographical position of the ship. Due to these reasons, the influences of the salinity, the alkalinity and the temperature on the absorption of SO_2 into seawater have been examined in a small packed column in the laboratory. Additionally, the influences of different salt compositions on the SO_2-absorption are investigated. A few selected experiments are repeated at a technical pilot plant (DN 450) to ensure the transferability of the experiments gained in laboratory scale to industrial scale. The presentation outlines the results of the experiments in the laboratory and at the technical plant.

444047 Utilization of Choline Cholire with Levulinic Acid and Phenylacetic Acid Based Deep Eutectic Solvents for Efficient CO_2 Capture

Abstract:

Choline chloride ([Ch][Cl]) + levulinic acid [LA] and ([Ch][Cl] + phenylacetic acid [PAA] deep eutectic solvent (DES) is studied as a suitable material for CO2 capturing purposes. The most relevant physicochemical properties of this solvent are reported together with the CO_2 solubility as a function of temperature. A theoretical study using both Density Functional Theory and Molecular Dynamics approaches is carried out to analyze the properties of this fluid from the nanoscopic viewpoint and their relationship with the macroscopic behavior of the system and its ability for CO_2 capturing. The behavior of liquid – gas interface is also studied and its role on the CO_2 absorption mechanism is analyzed. The combined experimental plus theoretical reported approach leads to a complete picture of the behavior of this new sorbent with regard to CO_2, which together with its low pricing, and the suitable environmental and toxicological properties of this natural solvent, lead to a promising candidate for CO2 capturing technological applications. Characterization techniques such as TGA, FTIR, viscosity, density, pH, and thermal conductivity analysis have been performed. Both in-situ FTIR and gas solubility measurements have also indicated the suitability of this material for CO_2 and other gases solubility. Rubotherm® sorption apparatus was employed for very accurate gas solubility experiments up to 30 bars and at 25 °C and 50 °C. Initial findings are reported in this communication and CO_2 solubility in [Ch][Cl]:[LA] was found to be as 2.12 mmol/g and [Ch][Cl]:[PAA] as 2.28 mmol/g. Moreover, corrosive behavior of the DES system have been experimented and compared with the most common amine based system and it has been observed that DES system has order of magnitude less corrosive behavior than the classical MEA CO_2 absorbent system.

443740 Integrated Performance Evaluation of High-Gravity Carbonation Process for Carbon Capture and Utilization: Exemplified By Petrochemical Industry in Taiwan

Abstract:

In this study, an integrated approach to establishing a waste-to-resource supply chain among different industries was developed for CO_2 capture, wastewater neutralization and product utilization via a high-gravity carbonation (HiGCarb) process, i.e., rotating packed bed (RPB). According to our previous experience, a high CO_2 capture efficiency (i.e., >98%) can be achieved by utilizing basic oxygen furnace slag (BOFS) and cold-rolling wastewater (CRW) via the HiGCarb process with a relatively short reaction time at ambient temperature and pressure. The objectives of this study were (1) to evaluate the resource potential of alkaline solid wastes in Taiwan for CO_2 capture and utilization using the HiGCarb process in Taiwan, (2) to quantify the pollution emissions and environmental impacts of the HiGCarb process from the cradle-to-gate life-cycle, (3) to calculate the water, carbon, and land footprints of the HiGCarb process, and (4) to establish the techno-economic analysis (TEA) of the HiGCarb process at an industrial-plant scale. The alkaline wastes were found to be successfully carbonated with CO_2 in the HiGCarb process, where calcite ($CaCO_3$) was identified as the main product. Since the reacted product (e.g., carbonated solid wastes) can be used as supplementary cementitious materials, CO_2 emissions from the cement industry can be also avoided if a green waste-to-resource supply chain between the steelmaking and cement industries is established. It was concluded that an integrated approach to the proper treatment of alkaline wastes that permanently fixes CO_2 from the petrochemical industry while producing valuable supplementary cementitious materials for the cement industry can be achieved via the HiGCarb process.

444172 Conversion of Synthesis Gas into Ethanol Using Homogeneous Catalysts

Abstract:

Before production of fuel ethanol became dominated by fermentation routes, many of the large oil and chemical companies utilized synthesis gas produced from natural gas and patented methods for conversion of H_2 + CO into ethanol using *homogeneous* catalysis. They often employed metals long used in Fischer-Tropsch synthesis, because of the ability of these metals to catalyze CO-transfer and CO-insertion reactions. By a judicious choice of ligands and reactor residence time, chain lengths of the Fischer-Tropsch products could be limited to the lower molecular-weight alcohols. Ongoing work at TDA Research, Inc. demonstrates that a wide variety of transition-metal oxides are reduced by the H_2 + CO of high-pressure synthesis gas into carbonyl anions and anionic metal-carbonyl hydrides that remain stable at ambient pressure and temperature. In general, the TDA carbonyl anions catalyze CO-transfer reactions; hydrides transfer hydrogen for reduction reactions, and carbonyl hydrides transfer *both* CO and hydrogen. Rate to ethanol exceeds 0.26 g ethanol/g metal catalyst/h, which on a metals basis, is equivalent to the best heterogeneous catalysts reported, but with the homogeneous system showing much higher selectivity, especially in minimizing low-value CH_4.

Acknowledgement: U.S. Department of Energy, Office of Science, Grant Award: DE-SC0004376.

16AIChE Spring Meeting, Houston, TX
12TH GLOBAL CONGRESS ON PROCESS SAFETY

Exploring Options for Mercury Removal Process within an Existing Condensate Stabilization Unit

Mahdi Nouri & Eberhard Lucke, CH2M OG&C, Houston, TX
Seyi Odueyungbo, Chevron ETC, Richmond, CA

ch2m: Chevron

Agenda

16AIChE Spring Meeting, Houston, TX
12TH GLOBAL CONGRESS ON PROCESS SAFETY

- Overall Block Flow Diagram
- Adsorbent Process Fundamentals
- Design Basis
- Condensate Stabilization Overview
- MRU Location
- Study Methodology for Data Screening
- Economic Analysis
- Conclusions
- Q&A

ch2m: Chevron

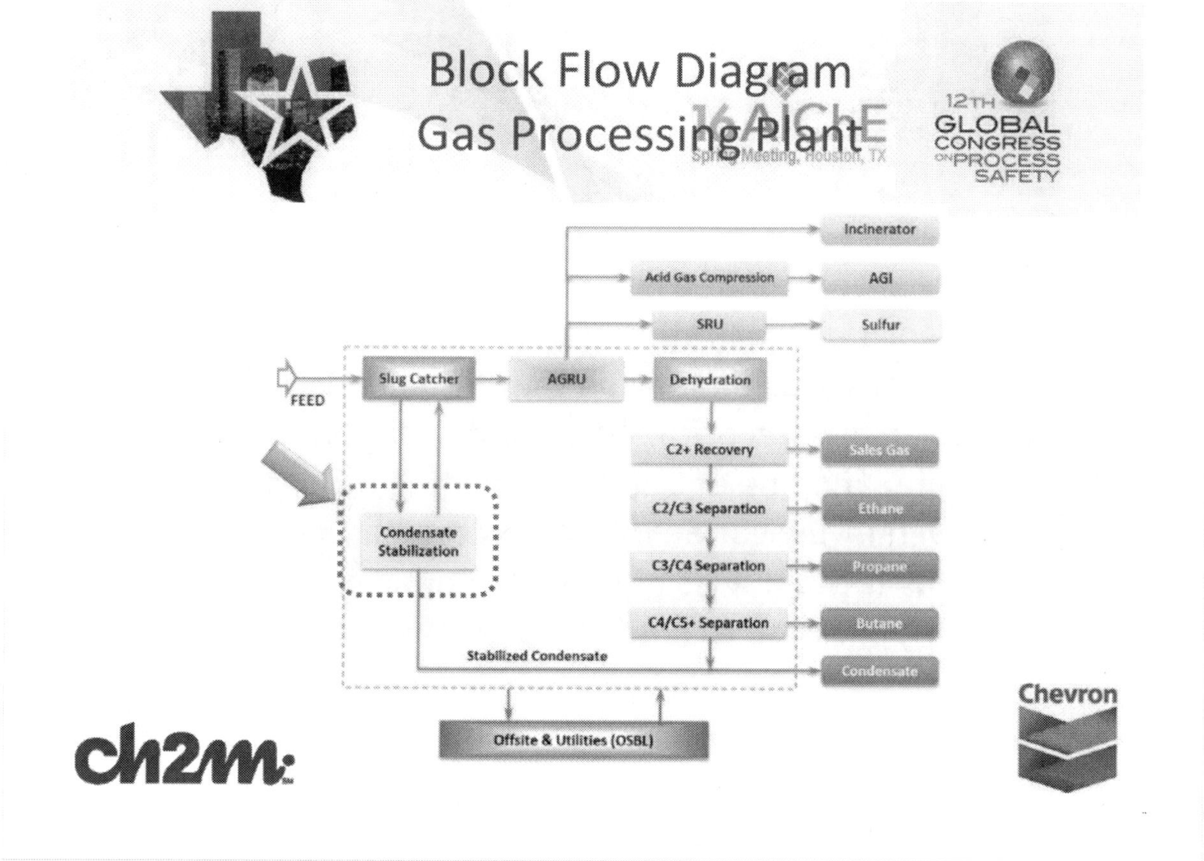
Block Flow Diagram
Gas Processing Plant
12TH GLOBAL CONGRESS ON PROCESS SAFETY
Incinerator
Acid Gas Compression
AGI
SRU
Sulfur
FEED
Slug Catcher
AGRU
Dehydration
C2+ Recovery
Sales Gas
C2/C3 Separation
Ethane
Condensate Stabilization
C3/C4 Separation
Propane
C4/C5+ Separation
Butane
Stabilized Condensate
Condensate
Offsite & Utilities (OSBL)
ch2m:
Chevron

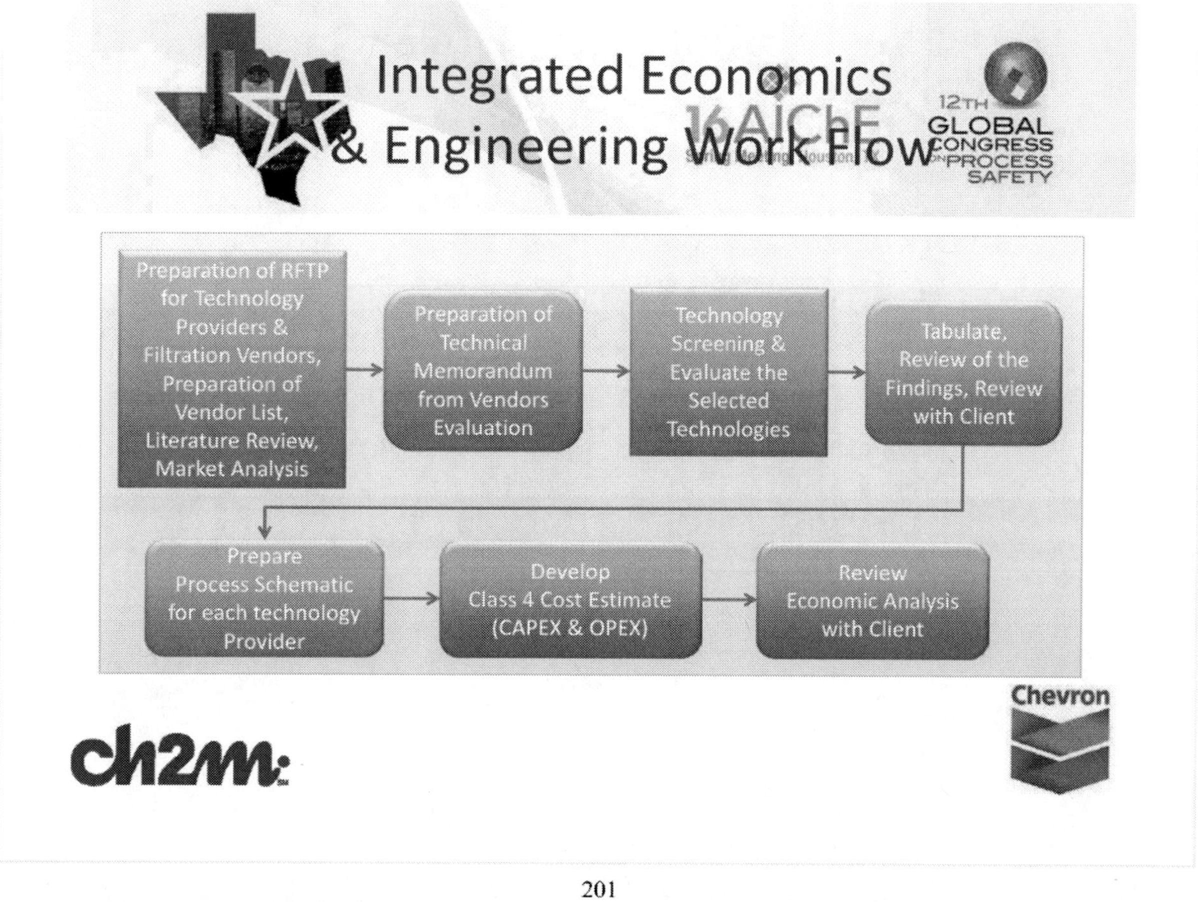
Integrated Economics & Engineering Work Flow
12TH GLOBAL CONGRESS ON PROCESS SAFETY
Preparation of RFTP for Technology Providers & Filtration Vendors, Preparation of Vendor List, Literature Review, Market Analysis
Preparation of Technical Memorandum from Vendors Evaluation
Technology Screening & Evaluate the Selected Technologies
Tabulate, Review of the Findings, Review with Client
Prepare Process Schematic for each technology Provider
Develop Class 4 Cost Estimate (CAPEX & OPEX)
Review Economic Analysis with Client
ch2m:
Chevron

Project Goals

16AIChE Spring Meeting, Houston, TX

12TH GLOBAL CONGRESS ON PROCESS SAFETY

- To review and evaluate mercury removal technologies for removal of elemental mercury from hydrocarbon condensate streams within an existing condensate stabilization unit
- To establish technology limitations and identify optimal location of the MRU and filtration package
- Develop and present optimal treating process through investigation into technical feasibility of various technologies and an economic evaluation of their application to liquid hydrocarbon stream.

ch2m: Chevron

Mercury in Natural Gas & Condensate

16AIChE Spring Meeting, Houston, TX

12TH GLOBAL CONGRESS ON PROCESS SAFETY

Location	Gas (mg/m3)	Liquids (mg/kg)
Europe	100 - 150	-
South America	50 - 120	50 - 100
Gulf of Thailand	100 - 400	400 - 1200
Africa	80 - 100	500 - 1000
Gulf of Mexico (USA)	0.02 - 0.4	-
Overthrust Belt (USA)	5 - 15	1 - 5
North Africa	50 - 80	20 - 50

ch2m: Chevron

Adsorbent Process Fundamentals

- Sulfur Impregnated Activated Carbon,

$$Hg + S \rightarrow HgS$$

- Metal Sulfides Dispersed in a Solid Carrier,

$$Hg + MSx \rightarrow MS(x-1) + HgS$$

- Molecular Sieve Technology

Design Basis

- Feed Actual Plant Data

Stream No. Description	<11> HC to Stabilizer Condensate Mixing Drum		<14> HC from Stabilizer Condensate Mixing Drum		<17> Condensate Stabilizer Bottom	<18> Stabilized Condensate Product
Phase	V	L	V	L	L	L
Temperature (C)	18.1		17.2		200.1	41
Pressure (Psia)	145		101.5		89.92	88.03
Normal Flowrate (m3/hr)	202.8	75.21	46.14	74.56	88.69	71.26
Density (kg/m3)		783.9		784.7	619.6	771.1

- Product Specification: <u>>10 ppbwt</u> of Mercury Content

Condensate Stabilization Process Flow Diagram

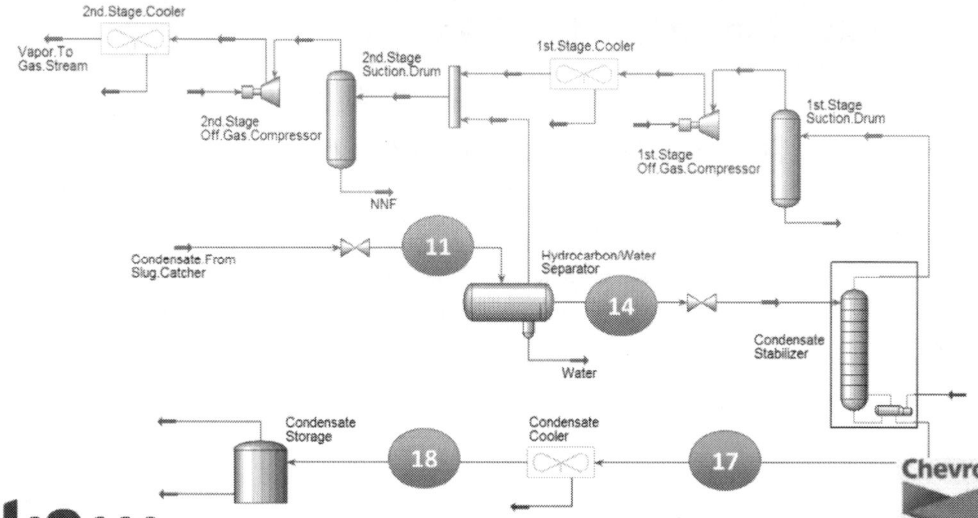

MRU Process Scheme and Proposed Location

F-100

Particulate Filter

V-100 A/B

Mercury Removal Beds

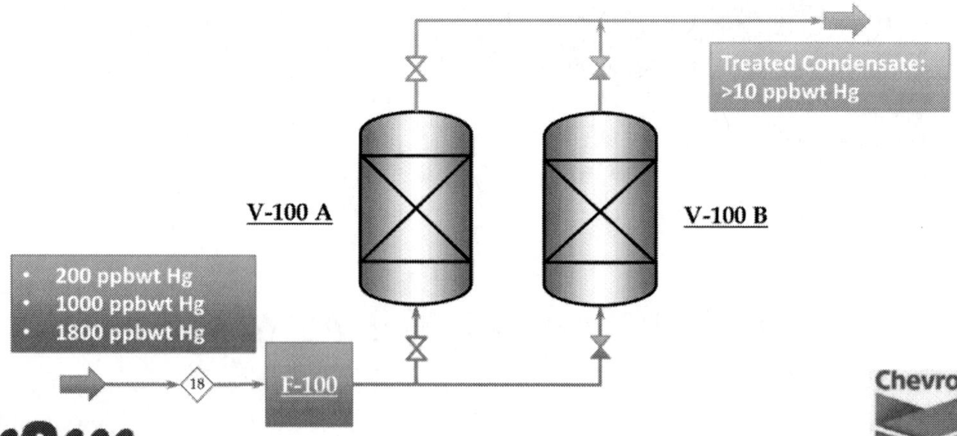

Study Methodology for Data Screening

12TH GLOBAL CONGRESS ON PROCESS SAFETY

DESCRIPTION	CASE 1	CASE 2	CASE 3	CASE 5
Hg inlet concentration (ppbwt)	1800	1000	200	1800
Hg outlet concentration (ppbwt)	< 10	< 10	< 10	< 10
Location of bed	--	--	--	--
Mercury removal bed				
- Flow direction	--	--	--	--
- Bed configuration	Single	Single	Single	Single
- Diameter (m)	--	--	--	--
- Height (m)	--	--	--	--
- Volume (m3) /bed	--	--	--	--
- Mass (kg) /bed	--	--	--	--
- Pressure drop (Psi)	--	--	--	--
- Design life of adsorbent	1yr	1yr	1yr	4yr

ch2m

Chevron

Mercury Removal Bed Mass of Adsorbent

12TH GLOBAL CONGRESS ON PROCESS SAFETY

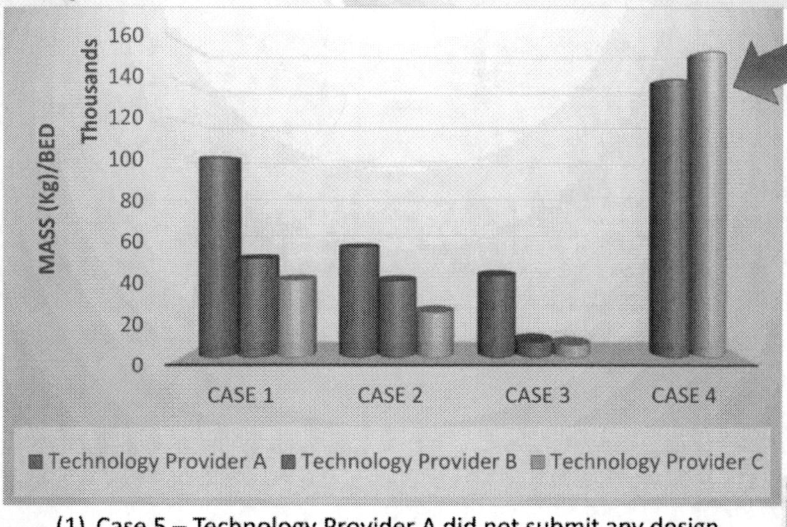

(1) Case 5 – Technology Provider A did not submit any design info (bed size limitations), others proposed 2xparallel beds.

ch2m

Chevron

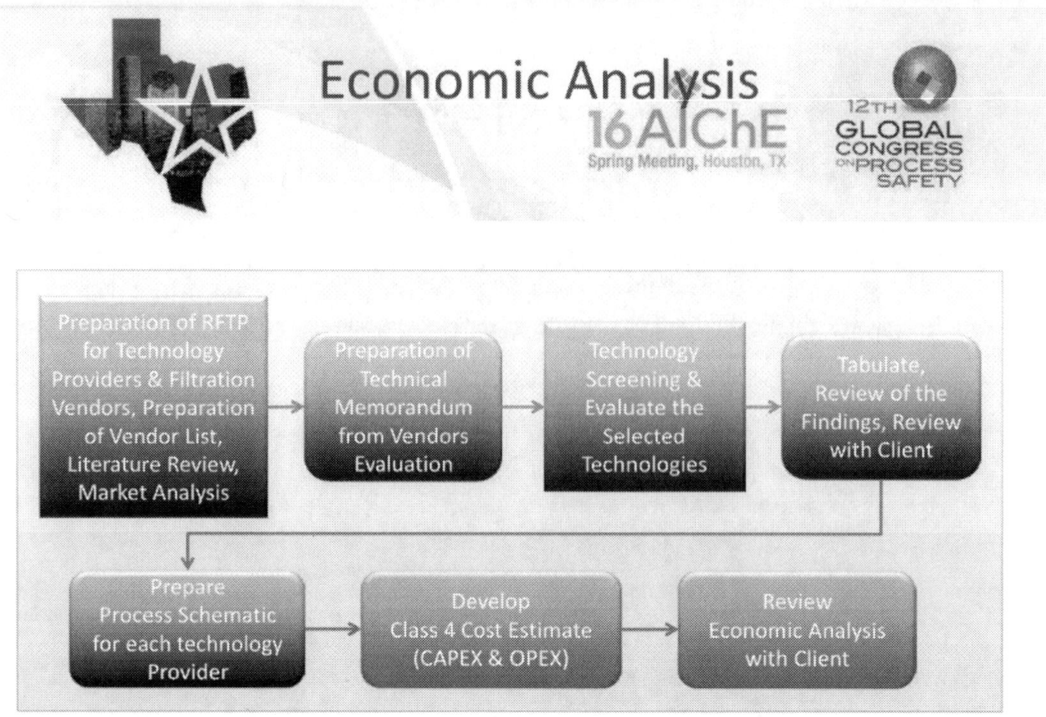
Economic Analysis
16AIChE
Spring Meeting, Houston, TX
12TH GLOBAL CONGRESS ON PROCESS SAFETY
Preparation of RFTP for Technology Providers & Filtration Vendors, Preparation of Vendor List, Literature Review, Market Analysis
Preparation of Technical Memorandum from Vendors Evaluation
Technology Screening & Evaluate the Selected Technologies
Tabulate, Review of the Findings, Review with Client
Prepare Process Schematic for each technology Provider
Develop Class 4 Cost Estimate (CAPEX & OPEX)
Review Economic Analysis with Client

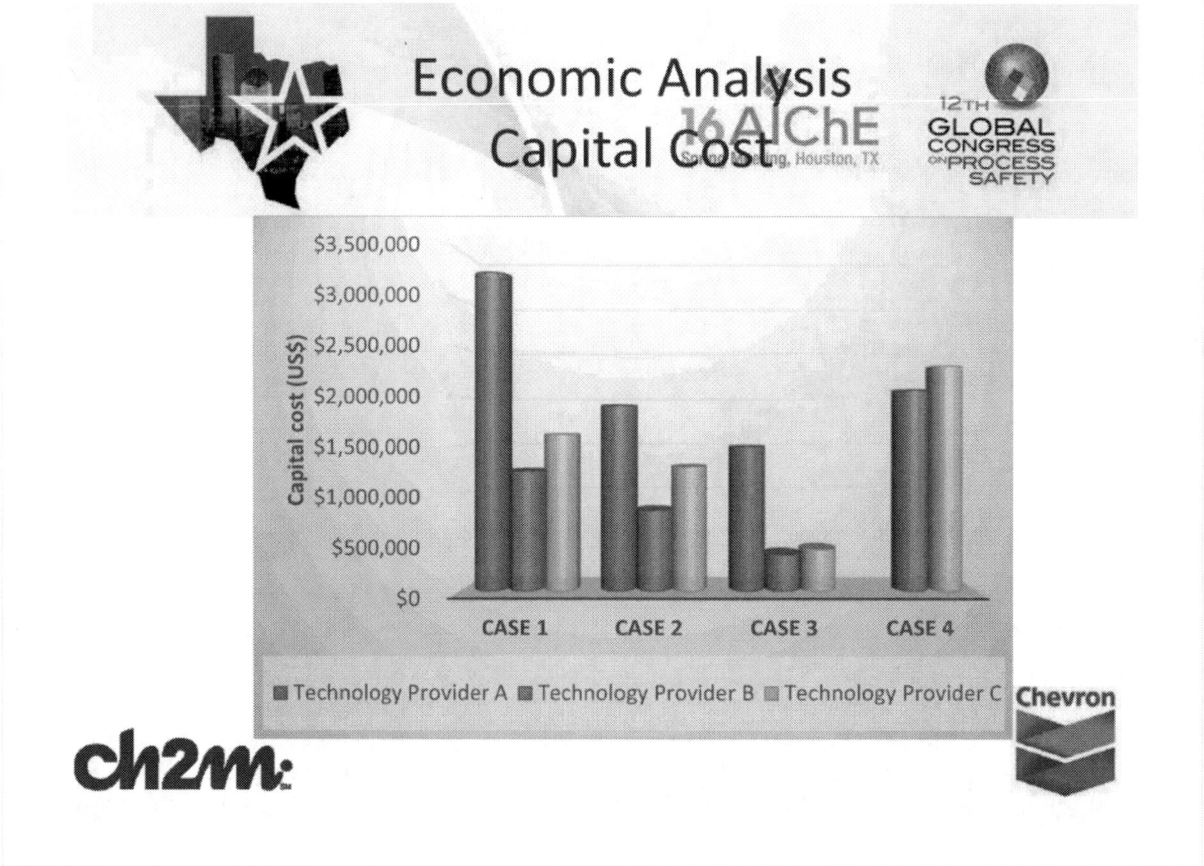
Economic Analysis
Capital Cost
16AIChE
Spring Meeting, Houston, TX
12TH GLOBAL CONGRESS ON PROCESS SAFETY
Capital cost (US$)
$3,500,000
$3,000,000
$2,500,000
$2,000,000
$1,500,000
$1,000,000
$500,000
$0
CASE 1
CASE 2
CASE 3
CASE 4
Technology Provider A
Technology Provider B
Technology Provider C
ch2m
Chevron

Economic Analysis Operating Cost

16AIChE Spring Meeting, Houston, TX

12TH GLOBAL CONGRESS ON PROCESS SAFETY

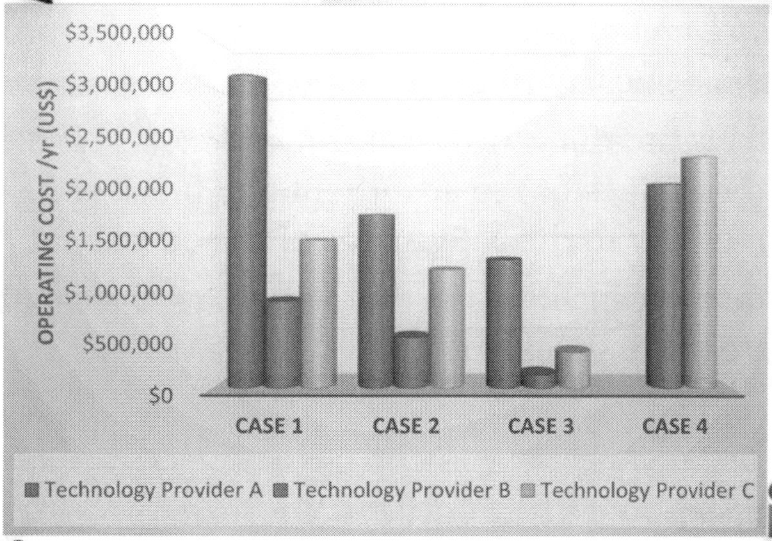

ch2m:

Chevron

Conclusion

16AIChE Spring Meeting, Houston, TX

12TH GLOBAL CONGRESS ON PROCESS SAFETY

- The speciation of mercury in hydrocarbon liquids was simplified to elemental mercury and particulate mercury of varying concentrations. The elemental and particulate allowed for the development of species specific mercury removal strategies in the future.

Chevron

ch2m:

Conclusion (Cont'd)

16AIChE Spring Meeting, Houston, TX

12TH GLOBAL CONGRESS ON PROCESS SAFETY

- The distillation range of the hydrocarbon condensate stream indicates the presence of hydrocarbon molecules up to C12+ in size.
- Long paraffinic hydrocarbon molecules tend to plug the surface of the adsorbent and block access of mercury to the active sites of the adsorbent.
- three licensors claim that they have the capability to treat the specified liquid HC stream

ch2m: Chevron

Conclusion (Cont'd)

16AIChE Spring Meeting, Houston, TX

12TH GLOBAL CONGRESS ON PROCESS SAFETY

- TP-A had operational experience with a heavy HC condensate, similar in composition and normal boiling point (NBP) to the basis for this study.
- TP-B shared a reference list; however, they did not provide their experience on C12+ compounds.
- TP-C submitted a technical proposal, they indicated a lack of confidence in their proposal and stated that they would be unable to provide standard performance guarantees for treating the specified condensate stream.

ch2m: Chevron

Conclusion (Cont'd) Location of MRU

- Several locations (stream 11, 14, 17, and 18) were evaluated within an existing condensate stabilization unit.
- Stream #18 (stabilized condensate product) was considered to be the most suitable location for the MRU package due the sub-cooled nature of the stream.

Conclusion (Cont'd) Safety Issues

- Most government agencies classify equipment and materials contaminated by mercury as hazardous waste that needs to be disposed of properly by local regulations.
- TP-A and TP-B indicated that they offer waste disposal programs to their customers to assist in managing the entire life cycle of the adsorbent.
- TP-C does not offer any waste management.

Conclusion (Cont'd)
Summary

16AIChE Spring Meeting, Houston, TX

12TH GLOBAL CONGRESS ON PROCESS SAFETY

Summary of economic analysis for MRU at worst operating case condition for achieving the product specification are:

- Mercury inlet concentration of 1800 ppbwt,
- Single bed configuration (2x100% beds),
- 1 year design life of adsorbent (Case 1)

ch2m:

Chevron

Conclusion (Cont'd)
CAPEX Selected Option

16AIChE Spring Meeting, Houston, TX

12TH GLOBAL CONGRESS ON PROCESS SAFETY

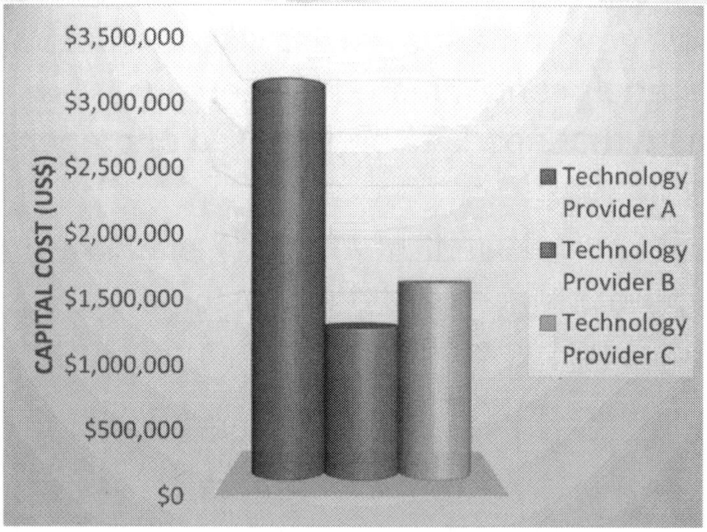

ch2m:

Chevron

Conclusion (Cont'd)
OPEX Selected Option

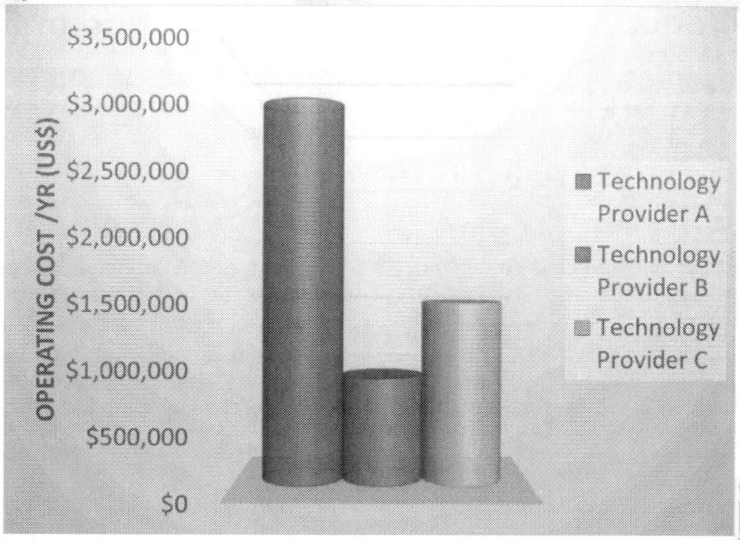

INHA UNIVERSITY

Optimization and Economic Evaluation of Industrial Wastewater Treatment Using Electro-oxidation Process

Hongmin Lee, Sangsun Lee, Yoonkyung Kim, Dohyung Lee, Sunghyeon Baeck, Sungwon Hwang*

Graduate School of Chemistry and Chemical Engineering, Inha University

Corresponding author* : sungwon.hwang@inha.ac.kr

2016. 04. 13

Contents

1. Introduction

1. Introduction

- SK Energy wastewater treatment process

  Wastewater collection pond & tanks → Oil separator → Flocculation/Coagulation tank → DAF→ Bio-reactor → Clarifier → Sand filter → Activated carbon filter → Effluent pond

- Physical/Chemical + Biological + High class Treatment
- COD_{mn} removal 200 mg/L → 10 mg/L

1. Introduction (Con't)

- Wastewater discharged from refinery or petrochemical plant contains high concentration of organic pollutants and toxic compounds during turnaround period.
- Conventional wastewater treatment system is generally ineffective in removing non-degradable organic. It requires additional pre-treatment system.

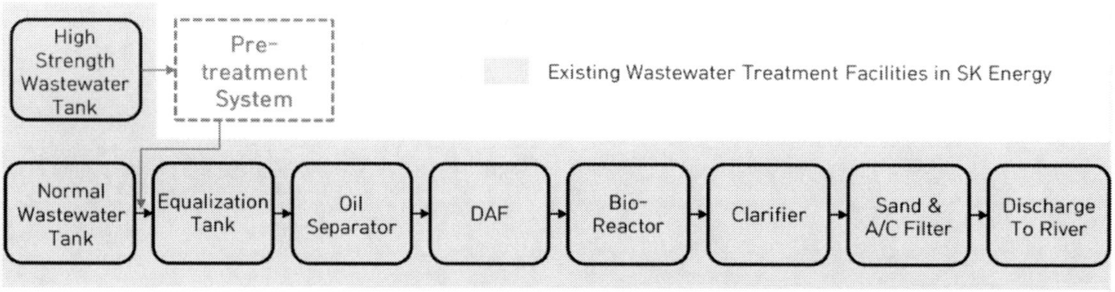

1. Introduction (Con't)

- Available Technologies

  ① **Incineration**
  - High temperature (820-1100℃)
  - Direct combustion with excess oxygen from air in flame
  - Applied mainly for concentrated wastewater with COD > 100 g/L

  ② **Wet Air Oxidation (WAO) Process**
  - Elevated temperature (250-300℃) and high pressure (100-150bar)
  - Usually Cu^{2+} is used as a catalyst.
  - The optimal COD is in the domain : 50 g/L > COD > 15 g/L

  ③ **Oxidation with strong oxidants**
  - The oxidation of organic pollutants with strong oxidants (H_2O_2, O_3) at room temperature
  - The oxidation takes place in the presence of catalyst and UV radiation
  - Treatment of dilute wastewater with COD < 5 g/L

2. Fundamental of Electro-oxidation

2. Fundamental of Electro-oxidation

1. Electro-oxidation reaction mechanism

$$S + H_2O \rightarrow S(\cdot OH) + H^+ + e^-$$

$$S(\cdot OH) + R \rightarrow S + Mineralization\ products + H^+ + e^-$$

$$S(\cdot OH) \rightarrow S + \frac{1}{2}O_2 + H^+ + e^-$$

- Items to be considered for electro-oxidation reaction mechanism :
  - Organic oxidation reaction, OER (Oxygen Evolution Reaction)
  - Adsorption enthalpy of S-OH

2. Fundamental of Electro-oxidation (Con't)

2. Oxidation power of the anode material

Electrode	Oxidation potential(V)	Overpotential of O_2 evolution(V)	Adsorption enthalpy of S-OH	Oxidation power of the anode
RuO_2-TiO_2	1.4-1.7	0.18	↑	↓
IrO_2-Ta_2O_5	1.5-1.8	0.25		
Ti/Pt	1.7-1.9	0.3		
Ti/PbO_2	1.8-2.0	0.5		
Ti/SnO_2-Sb_2O_5	1.9-2.2	0.7		
Ti/BDD	2.2-2.6	1.3		

2. Fundamental of Electro-oxidation (Con't)

3. Fabrication of Ti/SnO_2-Sb_2O_5 electrode

① Spray pyrolysis

② Sol-gel process

③ Sputtering

④ Chemical vapor deposition (CVD)

⑤ Electron beam evaporation

⑥ Thermal oxidation

SEM images of Ti/SnO_2-Sb_2O_5 electrode

3. Results of the Experiment

- **Experimental condition**

- Wastewater composition (ppm)

COD	pH	H_2S	T-N
70,000	8~10	3,000	3,000

Electrode area	6 x 7.5 = 45cm^2
Anode	Ti/SnO_2-Sb_2O_5
Cathode	Hastelloy
Current density	80mA/cm^2
Electrolyte volume	450mL
Temperature	50℃
Time	9h

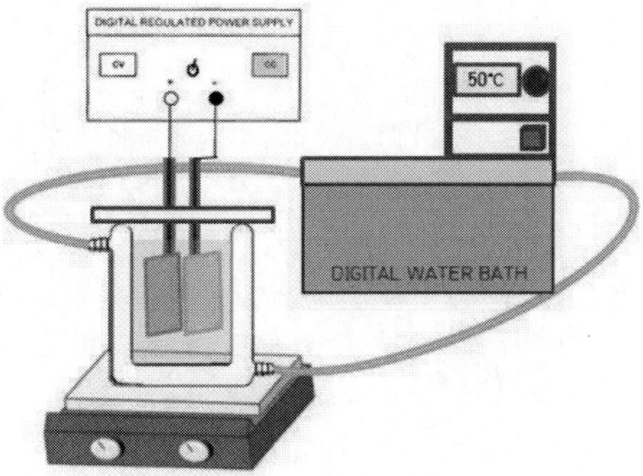

<Electro-oxidation Reactor >

3. Results of the Experiment (Con't)

- **Electrode preparation : Ti/SnO_2-Sb_2O_5**

1) Ti plate pre-treatment

Boiling DI water 30min → Boiling IPA 30min → Boiling 20vol.% HCl 5min → Sonication in DI water 10min → Boiling 10wt,% Oxalic acid 10min → Sonication in DI water 10min → Drying in oven at 70℃

2) Spray coating

(SnO_2 : Sb_2O_5 = 7:3, 8:2, 9:1)

$SnCl_2$ + $SbCl_3$ in ethanol → **Drying in oven at 70℃ 5min** → **Calcination at 400℃ 10min** → **Heat treatment at 500℃ 1hour**

Repeat 10 times

3. Results of the Experiment (Con't)

- Design and operating parameters

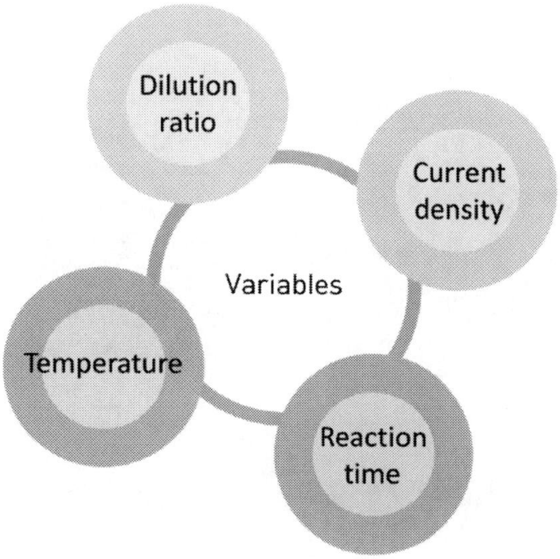

- Design the experiments using Central Composite Design (CCD)
- The experiment under various operating conditions
- Optimization of the variables
- Impact on economic evaluation such as payback period (PBP)

3. Results of the Experiment (Con't)

- Electro-oxidation experiments results

Anode	Ti/SnO_2-Sb_2O_5
Cathode	Hastelloy
Electrode area	6x7.5=45 cm^2
Electrolyte volume	450 mL
Time	9h
Initial COD concentration	70,000~80,000 ppm
Dilution ratio	0, 5, 10, 50
Current density	16, 48, 80 mA/cm^2
Temperature	40, 50, 60 ℃

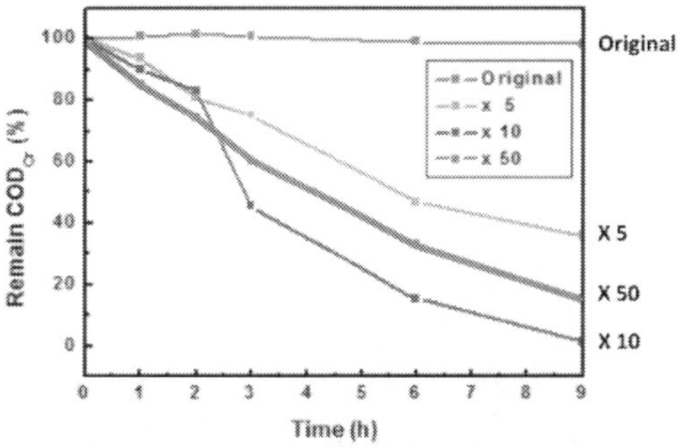

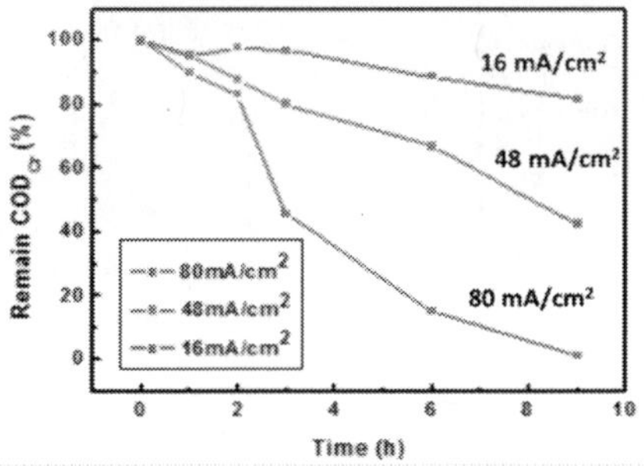

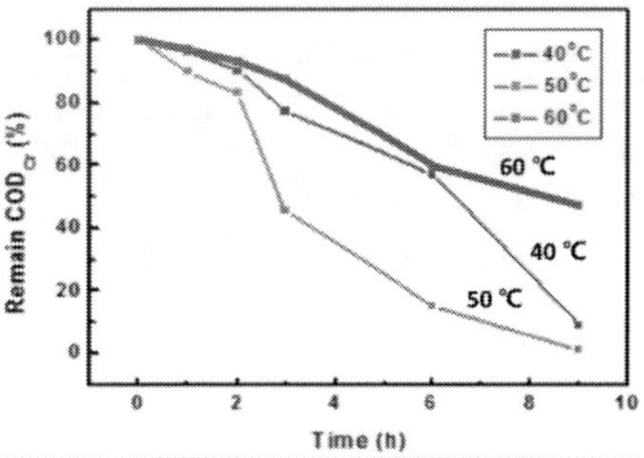

4. Results of the Modeling and Optimization

4. Results of the Modeling and Optimization

- Central Composite Design(CCD)

  ① Box and Wilson, 1951

  ② $y = b_0 + \sum b_i x_i + \sum b_i x_i^2 + \sum b_{ij} x_i x_j$

  - Linear terms in each of the variables$(x_1, x_2, \ldots, x_n)$
  - Squared terms in each of the variables$(x_1^2, x_2^2, \ldots, x_n^2)$
  - First order interaction terms for each paired combination

  ③ Relationship between codes and values

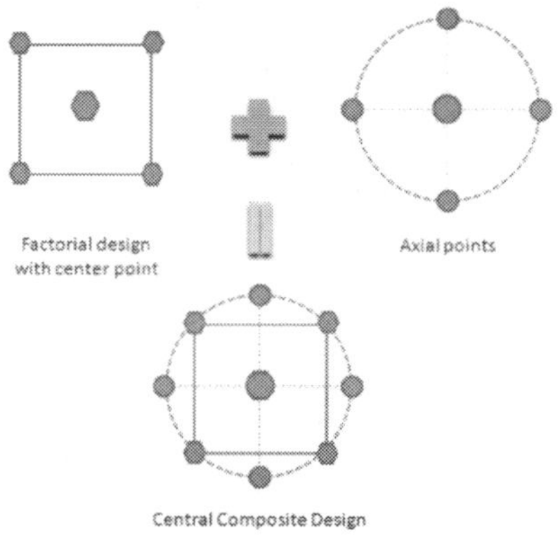

Codes	Values
-β	x_{min}
-1	$[(x_{max} + x_{min})/2] - [(x_{max} - x_{min})/2\alpha]$
0	$(x_{max} + x_{min})/2$
+1	$[(x_{max} + x_{min})/2] + [(x_{max} - x_{min})/2\alpha]$
+β	x_{max}

$\alpha = 1^{k/4}$

k = number of independent variables

4. Results of the Modeling and Optimization (Con't)

No.	Coded Level				COD removal(%)	
	x_1	x_2	x_3	x_4	Observed	Predicted
1	-1	1	0	-1	0	0
2	-1	1	0	0	1	5
3	-1	1	0	1	2	27
4	0	1	1	-1	6	6
5	0	1	0	0	43	43
6	0	1	0	1	64	69
7	1	1	0	-1	15	18
8	1	1	0	0	59	61
9	1	1	0	1	80	93
10	1	-1	0	-1	4	0
11	1	-1	0	0	9	14
12	1	-1	0	1	17	17
13	1	0	0	-1	4	6
14	1	0	0	0	29	36
15	1	0	0	1	59	53

No.	Coded Level				COD removal(%)	
	x_1	x_2	x_3	x_4	Observed	Predicted
16	1	1	0	-1	10	18
17	1	1	0	0	75	61
18	1	1	0	1	100	93
19	1	1	-1	-1	6	2
20	1	1	-1	0	38	54
21	1	1	-1	1	90	95
22	1	1	0	-1	10	18
23	1	1	0	0	75	61
24	1	1	0	1	100	93
25	1	1	1	-1	4	0
26	1	1	1	0	32	27
27	1	1	1	1	52	50
28	-1	-1	-1	-1	0	0
29	0	0	0	0	20	15

4. Results of the Modeling and Optimization (Con't)

- Central Composite Design(CCD)

$$CODremoval(\%) = 15.73 + 29.98x_1 + 25.48x_2 - 46.26x_3 + 18.79x_4 - 9.63x_1^2 + 1.78x_2^2 - 20.74x_3^2 - 5.98x_4^2 - 1.61x_1 x_2 - 21.27x_1x_3 + 4.79x_1 x_4 + 59.87x_2 x_3 + 14.07x_2 x_4 - 9x_3 x_4$$

b (15x1 double)

	1	2	3
1	15.7303		
2	29.9857		
3	25.4857		
4	-46.2689		
5	18.7982		
6	-9.6341		
7	1.7840		
8	-20.7478		
9	-5.9881		
10	-1.6117		
11	-21.2720		
12	4.7943		
13	59.8743		
14	14.0713		
15	-9.0000		

- Coefficient of determination

$$R^2 = \frac{SSR}{SST} = \frac{Sum\ of\ squares\ explained\ by\ reegression}{Total\ sum\ of\ squares}$$

$$= 0.92$$

4. Results of the Modeling and Optimization (Con't)

- Optimization of experimental results

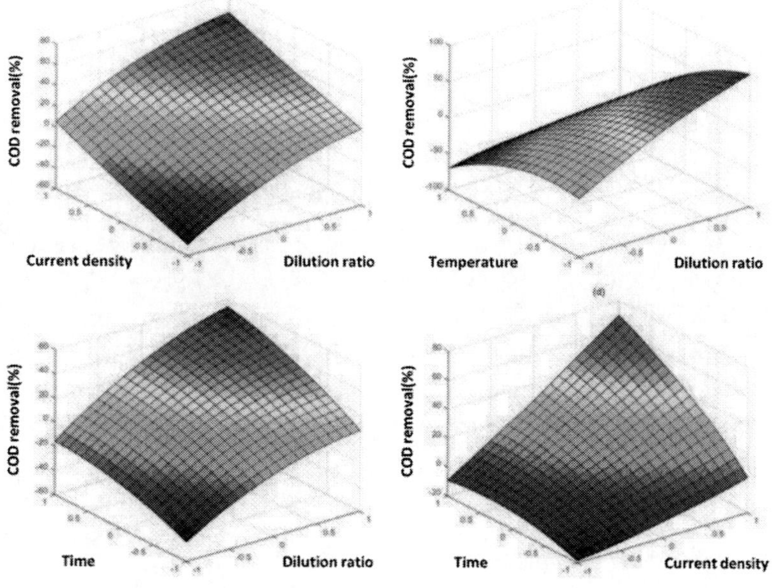

4. Results of the Modeling and Optimization (Con't)

- Optimization of experimental results
  - MATALB Optimization (Levenberg-Marquardt)

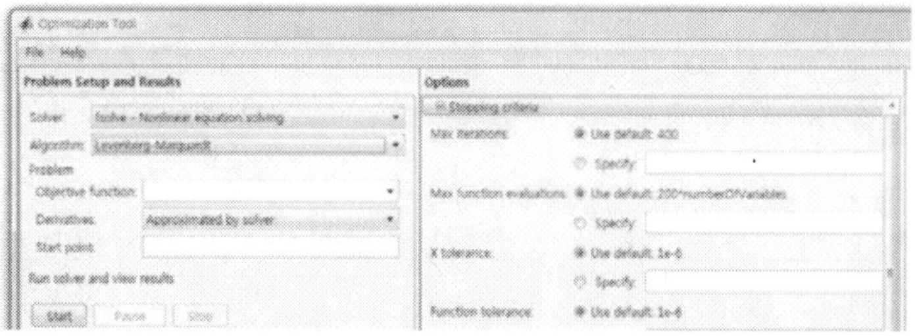

Objective : Y = 70 (COD removal efficiency)

Optimization variables : x_1, x_2, x_3, x_4

Subject to : $0 < x_1 < 16, 16 < x_2 < 80, 40 < x_3 < 50, 1 < x_4 < 9$

- Optimization results

① Dilution ratio(x_1) = 7.24 : 1

② Current density(x_2) = 57 mA/cm^2

③ Temperature(x_3) = 43.1 ℃

④ Reaction time(x_4) = 6.7 h

COD removal efficiency = 69.8%

4. Results of the Modeling and Optimization (Con't)

- Cost evaluation of electro-oxidation process
  - Main equipment : Electrochemical cell, the electrodes and the power supply
  - For industry applications, filter-press flow cell is widely recommended

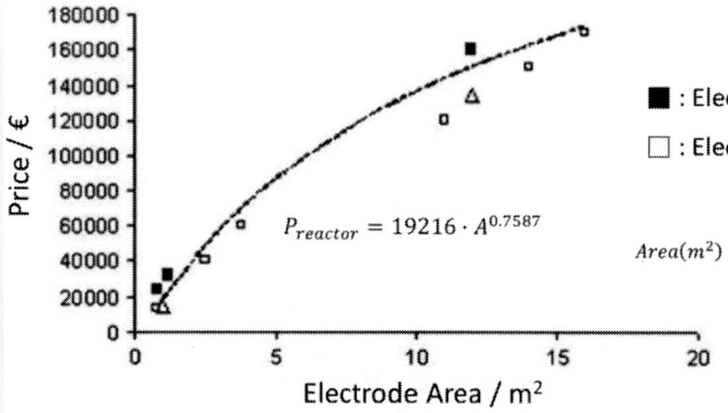

■ : Electrocell, Skjem, Denmark

□ : Electrochemistry Encyclopedia data

$$Area(m^2) = \frac{Q(A\,h\,dm^{-3}) \times 1000 \times flow\,rate(m^3\,d^{-1})}{j_{applied}(A\,m^{-2}) \times 24}$$

Canizares, Pablo, et al. *Journal of Environmental Management* 90.1 (2009): 410-420.

4. Results of the Modeling and Optimization (Con't)

- Cost evaluation of electro-oxidation process
  - Filter-press flow cell, electrode, electrical energy cost
  - Filter-press flow cell : \$ 1.2 million for a 100m^3 per day process
  - The price of electrode : 6,500 – 10,800 \$/m^2
  - Electrical energy cost : 0.08 \$/kWh
  - Operating cost and maintenance cost : \$ 540,000 per years

Estimated Payback Period (PBP) = 3.3 years

5. Conclusions and Future Works

5. Conclusions and future works

① Experiments

- CCD with 4 parameters and 3 level was used to design the experiments.
- The electrochemical oxidation of industrial wastewater with $Ti/SnO_2\text{-}Sb_2O_5$ electrodes
- Various experimental conditions : dilution ratio, current density, temperature and reaction time

② Modeling and Optimization

- The coefficients of CCD were estimated from the experimental results using MATLAB.
- MATLAB optimization was used to maximize the removal efficiency of COD under constrained condition.

  (Dilution ratio = 7.24 : 1, Current density = 57 mA/cm^2, Temperature = 43.1 ℃, Reaction time = 6.7 hours)

③ Cost Evaluation

- Filter-press flow cell was used to estimate the expense of the process (i.e. $1.2million for a 100m^3 per day).
- The estimated operating cost and maintenance cost : $540,000 per year
- Estimated PBP: 3.3 years

5. Conclusions and future works (Con't)

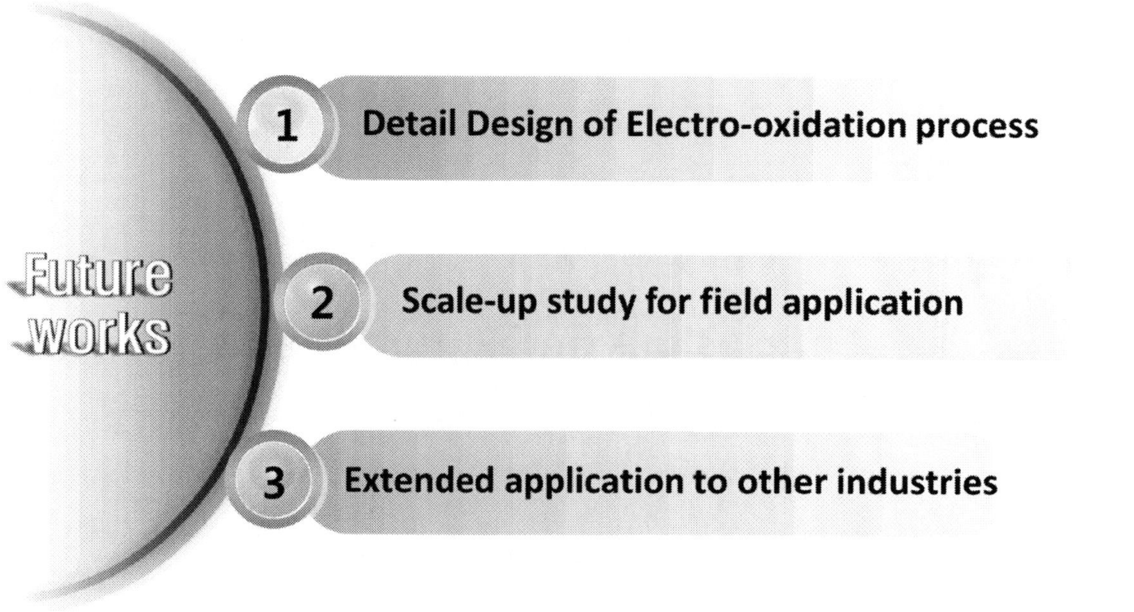

INHA UNIVERSITY

Q & A

This research was respectfully supported by Engineering Development Research Center (EDRC) funded by the Ministry of Trade, Industry & Energy (MOTIE). (No. N0000990)

"Development of Integrated Interactive Model for Subsea and Topside System to Evaluate the Process Design of Offshore Platform", funded by Ministry of Trade (project No. 10060099).

444092 Using Steel Slag for CO_2 Capture in Saturated Packed Bed

Abstract:

Carbon dioxide capture/sequestrate is the most promising options to reduce the CO_2 emissions into the atmosphere. Industrial mineral carbonation of alkaline wastes, an increasingly promising component of CO_2 capture/storage, may play an important role. Particularly the steelmaking slags are of interest because of their high calcium contents. In addition to CO_2 fixation, the carbonation of these waste materials also reduces the cost associated with handling the waste and minimizes the release of some harmful substances to the environment.

Considering the after mentioned advantages, a significant number of research initiatives have been undertaken at different organization in the world. These findings indicates that the prospect of carbonation of steel slag to become an economically-feasible CO_2 mitigation option. However, the successful implementation of this technology is likely to rely on the (i) availability of affordable and recyclable solvents/additives, (ii) accelerating and maximizing the CO_2 fixation by engineering the reactions involve of in order to achieve appreciable sequestration rate and (iii) detail economic analysis.

The proposed research mainly focused on enchantment of the CO_2 fixation using steel making slags by accelerating the reaction between CO_2 and the minerals in the slags. It is expected that the development and implementation of the proposed technology will create new opportunities both in the area of research and industrial applications to address both the CO_2 emission and steel management issues.

443922 New Approach to Using 3-D Dispersion Modeling for Accidental Toxic Gas Releases in Downtown Seoul

Abstract:

Chlorine is one of the most widely used chemicals in the world, especially for water purification plants near populated areas. So, if an accidental chlorine release occurs in a water purification plant, the damage would fatally spread to the residents and widely affect people.

In this paper, 3D model simulations are performed for a possible Cl2 leak scenario in the urban area of Seoul, South Korea where the population density is very high as well as numerous buildings exist near the plant in operation. FLACS is used to predict the dispersion results for the area whose size is around 5 by 3 km. In fact how precisely implementing 3D geometries is the key for successful 3D simulation. For that a methodology is suggested to rapidly implement geometry by importing CAD files provided by the South Korea National Spatial Information Clearinghouse, and processing it with another software. Finally, each FLACS simulation result was compared and analyzed with other dispersion modeling programs that do not reflect the real geometry.

NAVIGATING THROUGH REFINERY WASTEWATER TREATMENT PROCESSES AND PLANT CONFIGURATIONS.

Carlos D. Claros, P.E.,
CDM Smith, Houston, TX

Abstract

Petroleum refineries deal with a wide range of environmental compliance concerns encompassing wastewater discharges, air pollutant emissions and solids waste. Refineries must comply with stringent effluent discharge limitations which require extensive level of treatment and monitoring. Effluent discharges to surface water must pass whole effluent toxicity (WET) testing and meet very low levels of heavy metals and organics constituents. Some refineries discharge to Publicly-Owned Treatment Works (POTW), which impose less stringent limits, requiring varying levels of pretreatment and monitoring. Treatment processes used in refineries include oil water separators, dissolve air or gas floatation, activated sludge, nitrification-denitrification, selenium removal, filtration, adsorption and membrane processes. These processes generate residual sludge that needs to be dewatered and disposed. Control of hazardous air pollutants from treatment units is also an important consideration due to Volatile Organic Compound (VOC) emissions limitations. This paper presents an overview of typical wastewater treatment processes and configurations utilized in refineries.

Keywords: refinery, wastewater, compliance.

Refinery wastewater characteristics

Refinery wastewater consists of process wastewater, utility wastewater, and contaminated stormwater. Non-contaminated storm water runoff can be discharged directly to surface water provided it meets the regulatory limits. Utility wastewater consist of cooling tower blowdown, boiler blowdown, and residuals from water treatment. Sanitary wastewater is typically segregated and conveyed in a separate collection system.

Process wastewater originates from various process units including desalting, light oils, distillation, fractionation, alkylation, and reforming processes. Process discharges contain free and emulsified oil from leaks and spills, bottom tank draws; emulsions incident to chemical treatment; and condensates from distillate separators. Alkaline waste streams are discharged from the treatment of oil product with alkaline reagents and the sweetening of oil to remove mercaptans. Major wastewater streams are sour water, scrubber water from reformer catalyst regeneration, spent potassium hydroxide from alkylation, desalting wastewater, and caustic wash from isomerization. US EPA[1,2] has conducted surveys and studies for the various sources of wastewater generation and their characteristics.

In a typical refinery, process wastewater and utility wastewater are discharged to the process sewer collection system connected to refinery wastewater treatment plant (WWTP). During rain events, contaminated runoff also drains to the process sewer. The combined wastewater presents high variability in both flow and quality, so the WWTP must be designed to attenuate this variability to protect downstream treatment processes susceptible to upset and shock loads.

Table 1. Characteristics of Selected Process Discharges*.

Waste Stream Type	Condensates (Sour Waters)		Spent Caustic	Water Draws	Stripped Sour Water	Emulsion	Refinery Separator Inlet
Source of waste	*Low Sulfur Gas Separator*	*Light Oil Recovery*	*Naphtha Wash FCC*	*Slop Oil Treatment*	*Sour Water Stripper*	*Desalting Unit*	
Alkalinity, ppm	1518	69-13100	247900	--	--	739	0 - 92
Ammonia, ppm	500	3- 8500	-	--	30 - 100	2 – 100	--
Chemical Oxygen Demand (COD), ppm	1204	214-16,255	901200	22000 – 56000	600 – 1200	865-3031	--
Oil, ppm	3	6-230	1	4,900 – 10300	< 10	32-713	3900 – 6500
pH	8.5	5.0-9.2	13+	10 – 10.2	8	9.26	2.4 – 6.2
Phenol, ppm	0.06	0-213	309,300	--	< 200	4.1 - 100	0.5 – 7.9
Sulfide, ppm	600	10-5000	0-3380	--	1 - 10	1 – 100	--
Total Suspended Solids (TSS), ppm	-	14	678	60 - 940	< 10	< 500	131 – 678

* Adapted from Nemerow (1987) and IPIECA (2010)

Table 1 shows a list of conventional wastewater parameters for selected process discharges. Tank water draws, spent caustic, and desalter discharges have high organic loading and pH that could upset the wastewater treatment system if not properly managed within the treatment scheme. Condensates contain high concentrations of ammonia (NH_3) and sulfides, and are collected in the sour water tank for segregated treatment and reuse.

Other constituents in process discharges are chlorides, fluorides, mercaptans, heavy metals, selenium, volatile and semi-volatile organic compounds (VOCs and SVOCs), dioxins, and phenolic compounds. Toxic dioxin congeners (2,3,7,8 TCDD and 2,3,7,8 TCDF) have been found in treated effluent of activated sludge plants. The dioxins emanate from the scrubber water from reformer catalyst regenerator.

Effluent Limitations

Refineries effluent discharges are regulated by federal Effluent Limitation Guidelines (ELGs) set forth in 40 CFR Part 419 – Petroleum Refining Point Source Category. The U.S. EPA delegates the authority for issuance of facility discharge permits to the state and local agencies. The permit effluent limits are determined based on the following factors: refining operations category, crude processing capacity, age of plant, and whether the facility is a direct or indirect discharger. The wastewater treatment overview presented here-in focuses on the topping and cracking subcategories.

Table 2 shows the range of effluent concentration limits observed in recent permits issued to six U.S. refineries. Permitted parameters given in mass loading (e.g. metals) were converted to concentrations using the average annual flow. The direct dischargers included refineries with discharge of treated wastewater to an estuarine bay, a river, and a creek. The indirect dischargers included refineries with treatment plants including aerobic biological process that currently discharge to a POTW.

Table 2 – Range of Effluent Limits Observed in Recently Renewed Permits

Parameter	Direct Discharge (Surface Water)		Indirect Discharge (POTWs)
	Monthly Average	Daily Maximum	Daily Maximum
pH	6 - 9	6 - 9	5.0 – 12.5
Oil and Grease	10	10 -21	100
TSS	25 - 35	40 - 50	250 – 300
BOD	30 - 40	60 - 70	250 – 300
COD	200 - 300	450 – 550	--
Ammonia	3 - 20	35 – 45	100
Sulfides	0.15 – 0.2	0.005* – 0.45	3600
TPH	--	--	50
Chloride	--	300	--
Phenols	0.2	0.5	--
Benzene	--	--	0.05 - 0.5
Chromium	0.2 – 0.5	1.0	0.7 - 6.0
Hexavalent Chromiun	10 – 30	30 - 70	40
Mercury	--	0.001	0.005 - 0.2
Selenium	--	0.006*	0.05 – 7.5
Zinc	--	0.30*	2.0
Copper	0.01	0.025*	0.01
Lead	--	0.008*	0.05 – 0.15
Arsenic	--	0.20	0.2 – 1.6

* Limit based on dissolved concentrations

Federal effluent limitations and guidelines include metal limits only for total and hexavalent chromium; however, state and local permits have included other metal compounds based on surface water quality concerns. **Table 2** shows daily maximum concentrations for various metals including selenium, mercury, vanadium, and chlorides. Achieving such low limits of chlorides and selenium for surface water disposal may prove capital and O&M expensive.

Federal pretreatment standards are used by POTWs to establish effluent limits for categorical pollutants. In addition, the regulations allow POTWs to add pollutants or impose more stringent limits of regulated constituents when such limitations are necessary to implement the provisions of applicable Local or State regulations.

BWON Regulations

Benzene Waste Operations NESHAPs (BWON) regulations require emissions controls for waste streams with high concentrations of benzene (>10 ppmw). Waste streams from the crude unit desalter, sour water overflows, diesel coaslecers, and tank water draws commonly fall under this category. These are segregated and designated as BWON streams.

Refineries have been upgrading collection sumps and front-end treatment equipment such as American Petroleum Institute (API) separators, equalization tanks, and dissolved-air flotation (DAF) units with air tight covers and emissions controls to comply with BWON.

Treatment processes

The refinery wastewater treatment plant (WWTP) consist of the following general treatment process sequence:

- Oil and Solid Removal
    - Gravity Separation
    - Floatation
- Equalization
- Physico-chemical Treatment
- Biological Treatment
- Tertiary Treatment (Polishing)

Figure 1 shows the general treatment plant scheme used at refineries. Variations from this simplified process flow diagram are discussed in sections below. Segregated treatment is conducted for sour water and BWON discharges before entering the wastewater plant. IPIECA[3] describes these processes and technology options for each in more detail.

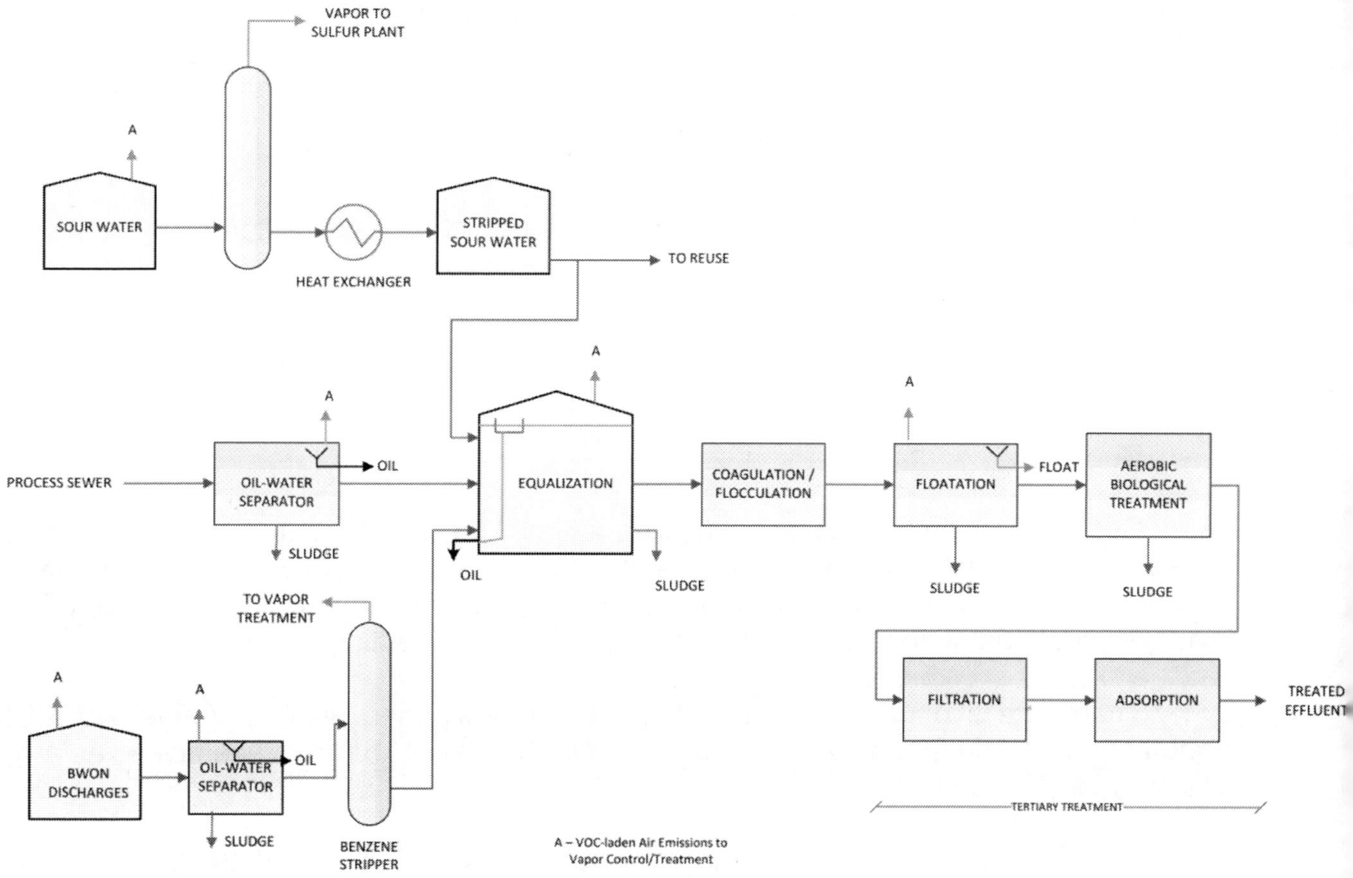

Figure 1 – General Refinery Wastewater Treatment Plant Process Flow

Oil and Solids Removal

The first step is to remove free-phase oil and suspended solids before the water is sent to equalization. API separators are the most commonly used gravity separation systems at the front end of the plant.

Some refineries use Coaslecing Plate Separation (CPS) units such as Corrugated Plate Interceptor (CPI), which is smaller in footprint than APIs. A common problem with CPI is that if not properly maintained they tend to foul and plug with light solids present in refinery wastewater, impairing separation efficiency. Demulsifiers may be used to facilitate oil separation in CPS units.

Secondary oil separation can be provided by a floatation unit following the API or CPI units, however this results in larger floatation units than if installed downstream of equalization.

Equalization

Equalization tanks attenuate wastewater flow and quality variability to protect downstream processes from fluctuations caused by high-strength waste discharges. These tanks are operated at a constant water level in flow through mode and keep available capacity for storm events.

Equalization tanks deprived of API separators are not common but are in place at some refineries where the oil layer is removed by skimming and/or lateral drain port withdrawals. Equalization tanks will accumulate sludge over time, reducing its effective volume, even more so for tanks not protected by API separators.

Physicochemical Treatment

Following equalization, the next step is the removal of emulsified oil by physicochemical treatment consisting of:

- pH adjustment
- Coagulation/Flocculation
- Dissolved or Induced Gas Floatation
- Chemical Oxidation

Coagulation and flocculation assists with oil de-emulsification and removal of heavy metals. Coagulation and flocculation uses ferric chloride and polymer to agglomerate metal precipitates into flocs that can be readily floated. In refineries, due to the presence of oil and low density solids, floatation is a more effective process than settling.

There are two general types of floatation processes used in refineries: Induced Air Floatation (IAF) and Dissolved Air Floatation (DAF) units. IAF removal efficiencies are typically less than for DAF units.

Some facilities use two stages of floatation, an IAF followed by DAF, where the IAF separates emulsified oil while the second stage removes metal by coagulation, flocculation and floatation. This configuration is robust and can produce a good effluent quality, however is more capital intensive.

Stringent BWON air regulations are an important driver for upgrades from DAF to DGF (G: nitrogen gas) units, because DGF units do not require a continual air sweep flow that DAF units need to keep VOC concentrations below lower explosive limits. The use of an air sweep in DAF creates a large volume of benzene-laden air that requires treatment by air pollution control devices.

Chemical oxidation reduces the effect of chelating compounds that may interfere with metal precipitation (e.g. cyanides or ammonia), oxidizes sulfides, and breaks oil emulsions. Chemical oxidation is an enhancement to physicochemical train that improves the efficiency of metal precipitation and helps with treatment of spent caustic discharges.

Segregated Treatment

Process discharges that pose significant pollutant loading are segregated in a separate equalization tank and/or pretreated before they are sent to the refinery WWTP.

BWON discharges

They contain high concentrations of benzene, which bring about emission concerns related to BWON regulations. These include desalter effluent, diesel coalescers, coker knock-out drum, and oil storage tank draws.

These waste streams are sent to a separate equalization tank equipped with oil skimmers for can be subsequently treated by steam or gas stripping to remove benzene and other VOCs. The VOC vapor stream is sent to an air pollution control device. The treated effluent is cooled and sent to the refinery WWTP.

Sour Water

Sour water consists of steam condensates generated in various refinery processes. Sour water carries low hydrocarbon content, but relatively high concentrations of hydrogen sulfide (H_2S) and NH_3, as shown in **Table 1**. This stream was also found to be the major source of selenium in refinery wastewater. Sour water stripping by either direct steam injection or reboiler steam removes H_2S and NH_3 from water but selenium remains in the water phase. The vapors go to the sulfur plant and the Sour Water Stripper (SWS) effluent is then reused in the desalter, flue gas scrubber, and hydrotreater make-up. The water that is not reused is sent to the refinery WWTP.

Sour water from delayed cokers and catalytic crackers also contain phenols and cyanides. These waters may be treated in a dedicated stripper to remove H_2S and NH_3, and then SWS effluent reused in the desalter where phenolic compounds can removed, reducing its loading to the refinery WWTP.

Selenium Treatment

Some refineries have been given very low selenium limits in the range of 5- 10 ppb during recent permit renewals. This has led refineries to assess selenium sources and implement upgrades for selenium treatment either at-source or end-of-pipe treatment, or as a combination of both. Feasible technologies are:

- SWS effluent treatment
  - Ion Exchange
  - Activated Sludge followed by Anoxic-Anaerobic Selenium Reduction
- End-of-pipe treatment
  - Iron co-precipitation
  - Specialty clay adsorption
  - Biological Anoxic-Anaerobic Selenium Reduction

Ion exchange has been successfully implemented as an at-source treatment of the SWS effluent but it has a high O&M and capital costs.

Anaerobic-anoxic biological treatment is an attractive alternative for selenium removal with low production of sludge, and is more cost competitive to ion exchange. This has been successfully piloted at a refinery as an end-of-pipe polishing step for indirect discharge, but full-scale implementation has been limited. This process causes an increase in WET toxicity due to sulfide formation, hence subsequent air oxidation would be required if discharging to surface water.

Biological Treatment

After free-phase and emulsified oil and metals have been removed, biological treatment removes dissolved organics and soluble oil. There are various biological treatment systems used at refineries.

Activated sludge operated in extended aeration mode has been reliably used for many years. Some systems have been retrofitted with support media for nitrification. Aerated lagoons have also been utilized over the years but refineries face regulatory pressures to phase them out due to more stringent VOC emissions regulations.

Aerobic Membrane Bioreactors (MBRs) are in the inception phase of implementation with several pilot tests showing promising results, and a few being taken to full-scale implementation. A variation of MBR uses suspended granular activated carbon (GAC) to help the reactor to better cope with shock loads of recalcitrant organics. Membrane materials available are PVDF and ceramic-type. The benefits of this technology are high organic removal efficiencies and nitrification/denitrification, which can help refineries meeting stringent toxic organics limits (e.g. benzene of less than 5 ppb) and nitrates associated with surface water discharge.

Tertiary treatment

Tertiary treatment (effluent polishing) consists of process units that remove TSS, recalcitrant trace organics and trace metals not removed in biological treatment. The typical tertiary treatment consists of filtration and media adsorption. Sand-anthracite media filtration are commonly used at refineries. Granular activated carbon (GAC) vessels are namely used for removal of trace organics although trace metal concentrations are also removed.

New pollutants of concern and/or more stringent effluent limitations are triggering WWTP upgrades at refineries with some occurring in this final treatment stage. For instance, new selenium and mercury limits are very low, and feasible treatment options include specialty adsorption media (modified clays), ion exchange, and anoxic biotreatment. Specialty GAC media can also be used for sulfides removal.

Some facilities discharging to surface water are being imposed low chlorides and TDS limits, which can be removed by Reverse Osmosis (RO). Alternatively, these constituents can be removed by at-source treatment of high TDS streams by desalination technologies (e.g. humidification-dehumidification).

Conclusions

New compounds of concern and more stringent effluent limitations have been affecting refineries at permit renewal for both direct and indirect discharge. Several treatment technologies and process configurations are available to upgrade and/or retrofit the existing plants to meet updated permit limits. There are new emerging technologies that do not have a track record in refineries but are worth considering during the feasibility analysis. Ultimately, the selected option for implementation needs to be pilot tested to confirm feasibility and demonstrate performance in achieving desired removal efficiencies.

Indirect discharge (POTWs) relieve facilities from more stringent surface water discharge limitations. However, potential upsets in the refinery WWTP may cause upsets in the POTW processes that could lead to disruptions of the refinery operation due to interruption of service while POTWs restore their plant operations. The refineries must protect against these events by identifying pollutant sources that could adversely affect both the onsite and POTW plants, and implement controls to minimize or manage these discharges.

Some refineries discharging to POTWs have treatment plants that provide extensive treatment, with some even including tertiary treatment. When considering treatment modifications and/or upgrades, treatment options that reuse treated effluent can prove beneficial to reduce pollutant discharge load, and in turn better meet effluent permit limitations.

As some refineries use more, or completely convert to, shale crude oil as feedstock, more studies are needed to understand whether there are changes in wastewater composition that could impact existing WWTP plant performance and/or discharge permit limitations.

References

1. US EPA (1996) - "Preliminary Data Summary for the Petroleum Refining Category", Office of Water, Engineering and Analysis Division.
2. US EPA (2004) – "Technical Support Document for the 2004 Effluent Guidelines Program Plan, Section 7 – Petroleum Refining" – EPA-821-R-04-014, Office of Water, Engineering and Analysis Division.
3. IPIECA (2010) – "Petroleum Refining Water/Wastewater Use and Management", Operations Best Practice Series.
4. Nemerow, L.N. (1987) – "Industrial Water Pollution: Origins, Characteristics, and Treatment; R.E. Krieger Pub.

443948 Refinery Wastewater Treatment for 'zero Liquid Discharge'

Abstract:

Entire wastewater from a large refinery was selectively treated for oil removal, then for BOD/COD removal and comingled with freshwater cooling tower blowdown. After equalization of the streams divalent cation were reduced by physical chemical treatment followed by multi-pass multistage Reverse Osmosis membrane process. Reverse Osmosis brine was further treated with falling film thermal evaporator. More than 95% of the water was recovered.

443539 New Refinery Wastewater Treatment Plant at Petron Bataan Refinery in the Philippines

Abstract:

Paper Is about the Design and Performance of a New Waste Water Treatment Plant with a Capacity of 700 m3/h for the Petron Bataan Refinery in the Philippines.Six Different Influents from the RMP-2 Refinery Are Sent to the WWTP. Treatment Included: Deoling - Biological Treatment and Solids Separation - Filtration - Disinfection - VOC and Deodorization

442942 It's All about the Money, Reducing Operating Costs in Heavy Metals Wastewater

Abstract:

It's all about the Money

Acid Mine Drainage (AMD) is naturally occurring water that continues draining from mines after they are closed. This water usually has high concentrations of heavy metals and a low pH. This water will flow for centuries and needs to be treated before it mixes with surface water.

Many of the mines that caused this drainage are closed so there is no responsible party. Many states and the EPA have set aside trust funds to treat this waste water from some of these sites. Existing mines are required to set up treatment plants and establish trust funds for operation of the treatment system before mine closure.

This worked for many years until interest rates dropped. The amount of interest on the trust fund was insufficient to operate the plants. The plants had to dip into the principal to pay operating costs. This paper outlines what was done to reduce costs and the problem that the changes caused.

443256 Managing Water Treatment Risks in Refining

Abstract:

Against the backdrop of global water scarcity, the industrial sector is the fastest growing source of water consumption. If current trends persist, by 2030, industrial water use will rise by 65 percent. Petroleum refiners and plants rank among the top industrial water users, consuming huge volumes of fresh water for cooling, steam generation and process needs.

Few would argue against the industry's need to reduce water and energy consumption, but accomplishing it is a challenge. Refiners are operating in an environment with wide variability in converting feedstock into high-quality products and where profitability hinges on keeping key process units running, controlling operating costs, and effectively meeting regulatory requirements. They are under pressure to expand their operating envelope, while addressing growing environmental concerns.

In this presentation, water treatment technologies, commercially available and under development, will be explored that facilitate the responsible and economical use of water in an increasingly demanding future. This presentation will highlight:

- Crude blend and desalter management while utilizing opportunity, or discounted, crudes in refining.
- Managing wastewater plant variation and risks associated with processing these crudes.
- Use of alternative influent waters to minimize fresh water consumption.

AUTHOR INDEX

AUTHOR INDEX

AIChE
120 Wall Street, FL 23
New York, NY 10005-4020

CURRAN ASSOCIATES INC.
proceedings.com

ISBN 978-1-5108-2493-5
9 781510 824935